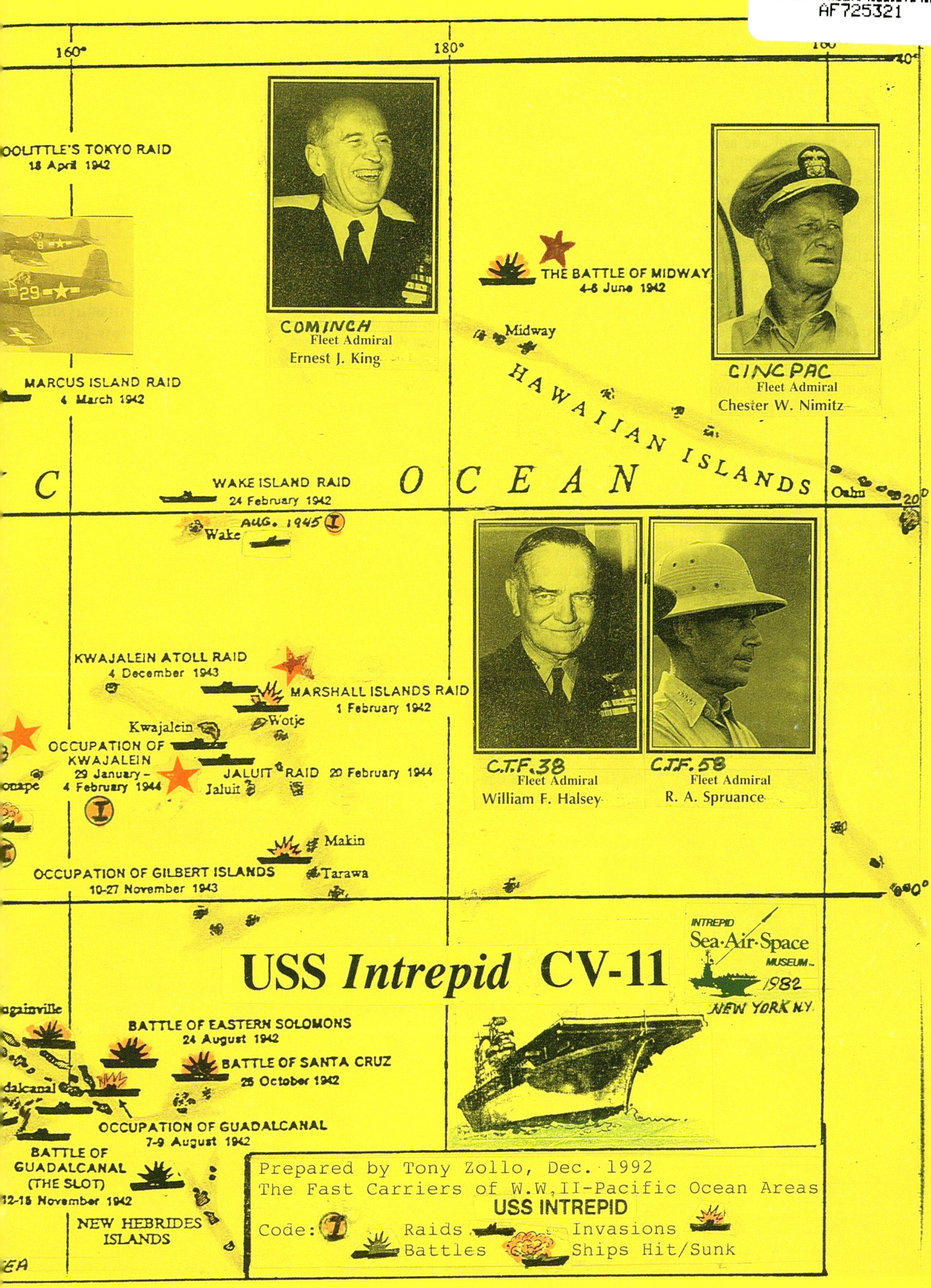

160°
180°
OOLITTLE'S TOKYO RAID
18 April 1942
COMINCH
Fleet Admiral
Ernest J. King
THE BATTLE OF MIDWAY
4-6 June 1942
Midway
HAWAIIAN ISLANDS
CINCPAC
Fleet Admiral
Chester W. Nimitz
MARCUS ISLAND RAID
4 March 1942
C
WAKE ISLAND RAID
24 February 1942
O C E A N
Oahu
20°
Wake
AUG. 1945
KWAJALEIN ATOLL RAID
4 December 1943
MARSHALL ISLANDS RAID
1 February 1942
Wotje
Kwajalein
OCCUPATION OF KWAJALEIN
29 January - 4 February 1944
JALUIT RAID 20 February 1944
Jaluit
C.T.F. 38
Fleet Admiral
William F. Halsey
C.T.F. 58
Fleet Admiral
R. A. Spruance
Makin
Tarawa
OCCUPATION OF GILBERT ISLANDS
10-27 November 1943
USS Intrepid CV-11
INTREPID
Sea·Air·Space
MUSEUM
1982
NEW YORK N.Y.
BATTLE OF EASTERN SOLOMONS
24 August 1942
BATTLE OF SANTA CRUZ
26 October 1942
OCCUPATION OF GUADALCANAL
7-9 August 1942
BATTLE OF GUADALCANAL (THE SLOT)
12-15 November 1942
NEW HEBRIDES ISLANDS
Prepared by Tony Zollo, Dec. 1992
The Fast Carriers of W.W.II-Pacific Ocean Areas
USS INTREPID
Code: Raids Invasions
Battles Ships Hit/Sunk

Warm Regards
Anthony F. Zello Sr.

USS Intrepid

CV-11
CVA-11
CVS-11

Turner Publishing Company
Publishers of Military History
P.O. Box 3101
Paducah, Kentucky 42002-3101

The materials were compiled and produced using available information; Turner Publishing Company and Mark A. Thompson regrets they cannot assume liability for errors or omissions.

Co-produced by Mark A. Thompson,
Independent Publishing Consultant
for Turner Publishing Company

Book Design: Elizabeth Dennis

Library of Congress Catalog
Card No. : 93-60655

ISBN: 1-56311-107-1

Printed in the United States of America

Limited Edition

***Front Cover Painting** was provided courtesy of the artist Kipp Solpwedel.*

***Back cover Photograph** was provided courtesy of the Intrepid Sea Air Space Museum*

Injured man waits for transport to sick bay, while fire is ongoing aft. (National Archives Photo)

UNDER 700 TONS
48
UNDER 700
10
ON THE GROUND - 183
FOX
BAKER
TARE
NO.
DN.
UP.
NO.
DN.
UP.

Intrepid heads to the Marshall Islands via the Hawaiian Islands. (National Archives Photo)

CONTENTS

God grants liberty only to those who love it and are always ready to guard and defend it.

Daniel Webster

PREFACE

They had a profound impact on history throughout their service life. The Essexes were the largest single class of capital warships ever built.

Born in crisis during the dark days of World War II, the Essex Class Fast Carriers were thoroughly battle tested and proven. Despite intense and continuous combat, none of these magnificent ships were lost because of enemy action. This was due in no small part to the men who manned them. The ships... the aircraft...the pilots...the crew...they were the best!

Now, fifty years later, almost all of the 24 venerable Essexes are gone, except for three sisterships which have been preserved as museums. This is the story about one of them, USS INTREPID, CV-11. It is written by one of her intrepid crew, former US Marine Anthony Zollo. His intense dedication to his ship is evident throughout this book. Tony kept a very accurate daily diary during his tour of duty aboard Intrepid and his extensively researched those actions which took place after he left.

Throughout her World War II service, Intrepid endured bombs, kamikazes and torpedoes. However, the Fighting "I" came back again and again. She was a graceful and splendid ship, crewed by courageous men. Sadly, some of them never came back. However, their memory is forever etched into Intrepid's heritage as she continues her second career as a museum ship in New York City.

After 31 years of proud service, Intrepid was retired in 1974. She had fought in history's largest naval battle and many campaigns during World War II. Her last combat was off Vietnam in 1968. In between conflicts, Intrepid was an invaluable asset as she supported the nation's international responsibilities and NASA's space program. She epitomized American power and prestige in the 1940s, 50s, 60s and 70s.

Even in retirement, Intrepid continued to serve. Shortly after being decommissioned, Intrepid was designated by Congress to be the US Navy and Marine Corps' Bicentennial Exposition Vessel at Philadelphia in 1976.

Two years later, she seemed to be doomed to be broken up for scrap as attempts to save her met with failure. Indeed, Intrepid would have been cut apart, suffering a fate similar to her sisterships, if it had not been for two very patriotic Americans.

Zachary and Elizabeth Fisher stepped in and led the campaign to preserve Intrepid. "For the generations yet to come...it is our hope that the young will never

Lt. Charles Devans (LDO) Landing Deck Officer on the Intrepid. (National Archives Photo)

have to endure war, but they must not forget that the price of freedom is never free. This battle scarred ship stands as our conscience in a fast changing world. Hundreds of young men died on her hallowed decks. This great country must never forget them..." said Zachary Fisher as he explained why he undertook such a mammoth project.

Mr. Fisher's enduring dedication has been personally inspiring to me. Now in her 50th year, Intrepid is bigger and better than ever, thanks to the Fishers and to the many Americans who have helped in many ways.

I am proud to be responsible for her care and preservation.

Lawrence Sowinski
Executive Director Intrepid Sea•Air•Space Museum

I wish to have no connection with any ship that does not sail fast, for I intend to go in Harm's Way.

John Paul Jones

DEDICATION

This is the story of the exploits of a mighty aircraft carrier during World War II (and its post—war duty) as seen through the eyes of a young Marine. The ship is the *U.S.S. Intrepid* and I was the young Marine (19 years old at the time), Anthony Zollo, 7th Marine Detachment, Captain's orderly and gunner, on board *Intrepid* from mid-1943 to early in 1945. The story of any great ship is really the hundreds — even thousands — of stories of her crew, officers and men. This is such a story. But this story goes beyond that of one ship, for wars at sea are not fought or won by a single vessel but by the combined fire power of a team. So we will write of other ships and engagements in this story. And the events that trigger a world war do not happen over one month or one year; the causes of World War II which began in Europe in 1939 can be traced back to the end of World War I in 1918. We will have reason to look at these events also.

Sailors tend to view their ship as a living, breathing thing, and always a lady — "treat *her* kindly and *she* will be kind to you"; "*She* is a strong and proud ship and *she* will protect you." So was the carrier *Intrepid* viewed by her crew. Her planes bombed, burned and gutted the enemy in strike after strike, and she in turn was bombed, rammed, burned and gutted *by* the enemy. But the mighty *Intrepid* always returned to the fight and she lives today. Since this great ship and its men fought World War II in the Pacific, the events chronicled here will be, primarily, America's war in the Pacific against Japan from 7 December 1941 to 2 September 1945. For those who were there and survived, these were the most momentous few years of their lives. We who came home are older now but those who did not come home will forever be young in our memory.

A. F. Zollo at the time of discharge 9 October, 1945.

The story of the *Intrepid* is the story in part of the Pacific war as I saw it, was a part of it, and as I wrote of it daily in my diary as it took place. My diary begins with my orders to go on board the carrier *Intrepid* as a member of the 7th Marine Detachment on August 16, 1943 and I would remain on board until the middle of February, 1945. I was present when the ship was torpedoed at Truk Island after the ship's strike on the Marshall Islands. The *Intrepid* then operated with Task Force 38 and for a time she would be the flag ship for Task Group 38.2 with Admiral Bogan in *Intrepid.* She will be hit many times and put out of action, but she always returned to take part in whatever strikes or invasions were on-going at the time.

She was present in the numerous sea battles in and around the Philippines for some eighty-four consecutive days. Not a day would go by but I made notes on what we were doing — where we were; the planes taking off on a strike or covering an invasion; the crack-ups on deck; the planes in the water out of fuel; the shot-up planes landing on the deck; the injured pilots and gunners in the returning planes; the kamikaze suicide attacks on the ship and the ones that got through and crashed into the ship; the remarks by the Admiral and the Captain and other officers — all jotted down in my diary. Being an orderly to the Captain and the Admiral, I was fortunate to hear and see much and I recorded it all.

When other ships were attacked or damaged or sunk, I made notes of it all. As a young 19 year old Marine I did not realize the extent of my notes at that time. After the war, for almost a year, I rewrote those notes, filling in the blank spaces and adding more information as I went along. I wrote of the weather, the sea, and the damage to the *Intrepid* and to other ships. I listed as best I could what ships were with us and what ships made up the task force of which we were a part. I wrote of the burial of our dead and, at times, of my own feelings. I wrote of the conditions which prevailed while in action, of the typhoons the fleet braved, and my trips ashore at Ulithi, our R & R (Rest and Rehabilitation) center. With all this writing done, I put the notes away for forty-seven years.

In 1991 I took this material out of my desk and began a massive typing effort to tell the story of the *Intrepid* — all of the travels and actions of this remarkable ship from her commissioning and shakedown trips to Trinidad and Maine to the Panama Canal, the Pacific, and war. Included in these writings are our actions at the Palau Island group, and the action and invasion of Saipan, Tinian and Guam. I write also of our raids on all the major Philippine islands: Leyte, Samar, Negros, Luzon and Mindanao. I write of our strikes on Formosa and Morotai, and on Kwajalein, Roi and Namur in the Marshall group. And I write of the *Intrepid* at Wake Island, Okinawa, Iwo Jima and Truk.

This story is dedicated to the KIAs and MIAs of *Intrepid's* men and pilots in two wars who did not come back, to the memory of my brother, C.F. Zollo of the 4th Marine Division, to my children Deborah and Tony, Jr., and most important to the memory of my dear departed wife and friend of forty-two years, Jennie. May they live in peace and rest in peace.

If there be any truer measure of a man than by what he does it must be what he gives.

Robert South

ACKNOWLEDGMENTS

My thanks to:

National Archives, Photographic Section, especially the Curator and still picture branches;

Washington Navy Yard, Naval Aviation Branch and ships history section.

Bernard F. Cavalcante, Head of Operational Archives, Naval Historical Center, Washington Navy Yard, for his kind words about me and for sending me the two daily log micro reels of *Intrepid* activity.

Richard W. Peuser, Archivist, National Archives and Record Administration, for his early on assistance and direction on the different naval records and photo sections.

A. J. Booth, Captain, U.S. Navy, Deputy Director of Naval History,

Department of the Navy, Office of the Chief of Naval Operations, for his direction to sources of certain naval information.

John F. Kerry, U.S. Senator, the first of my contacts, for his direction, and to his staff for their time and effort in obtaining certain military information.

Brian J. Donnely, Congressman, for his efforts in assisting me in obtaining my Marine Corps Service Records.

James E. Fahey, Archivist, Commonwealth of Massachusetts Military Division History and Museum, for his efforts in my behalf.

Patty M. Maddock, Naval Institute at Annapolis, Maryland, for her direction on whom to contact at the Institute.

Mark Gatlin, Acquisition Editor, Naval Institute Press, for his efforts in my behalf.

Frank Uhlig, Jr., Naval War College Review, for his assistance.

Evelyn M. Cherpak, Head of the Naval Historical Collection, Naval War College, Newport, Rhode Island, for her direction and assistance.

Sandy Scaccia and **Doreen Avila** for their typing of this book over the last two years.

Seventh Marine Detachment of *Intrepid*'s 1943 crew.

Paul Guertin of Dedham, Massachusetts of New England Home Movie Transfer for his assistance in photography and in making tapes for me.

Corrine Converse, Fast Print of Dedham, Massachusetts for the copying of hundreds of documents for me and for the many photo reproductions.

The courteous staff of the **Boston Public Library** for their assistance in the many hours of research of their collections.

The Boston newspapers, the **Post**, the **Globe**, and the **Herald** for certain news releases and stories of many years ago.

Frank Doria, on board Representative of the U.S.S. *Intrepid* Museum

Lawrence Sovinski, *Intrepid* Museum Director and Curator, for producing the Preface for this book.

Frank V. De Sesto, *Intrepid* Museum Curator of Photography, for the loan of Ship's photos.

Shirley J. Schultz, Chief of Documents and Volunteer coordinator on board *Intrepid* for the loan of photos.

The **Former Crew Members Association** Executive Board, for their photos and input.

Richard G. Masterson, for his information on Bombing 18, which flew off the *Intrepid* and his ongoing assistance on VB-18's history.

Mark A. Thompson, Publisher for Turner Publishing Company, for his expert suggestions and direction regarding my book.

Philip St. John, Ph.D., for his expert assistance in setting up this book and the editing of my material.

My son, for his work and storage of documents in my computer.

Vital Records of Randolph, Massachusetts, for developing hundreds of feet of micro film and transferring it to paper documents.

Department of the Navy, Comptrola Navy Imaging Command, for their photos sent to me.

Naval Historical Center, Ships History Section, Washington, D.C.

Naval Aviation History Branch, for their assistance.

***Intrepid*'s former Marines of 1943 and later** , for their input. Special thanks to **Colonel Lowel Keagy**, **Sergeant Dean Lesnett**, and **Corp's Ray Dow**. Special thanks to **George Cramer**, for his painting of *Intrepid* afire on 16 April, 1945.

To the **artist** of the painting of *Intrepid* under attack on 25 November, 1944.

Kipp Soldwedel for the loan of the original transparency of that painting so that I could use it for the book cover.

Dave Turner of Turner Publishing Company, for his time with me in Washington many months ago when he accepted my manuscript for publication.

Lee Fray, CAG-18 for his photos loaned to me.

All the **people** who have called me and written to me and loaned me photos for publication. Many Thanks!

Finally to my dear departed wife, who continually pushed me to do this story. I will always remember our 42 years together. May she rest in peace.

ABOUT THE AUTHOR

Anthony F Zollo, Sr.
Intrepid's 7th Marine Detachment, WW II

Anthony F. Zollo was born 5 March, 1924. He left high school (11th year) and went to work on ocean going tug boats prior to entering the Marines. (His brother preceded him by 11 months into the Marine Corps). Zollo entered the Marine Corps on 2 April, 1943. He trained at Parris Island and then attended Sea School at the Norfolk Navy Yard, VA. He was assigned to the carrier USS *Intrepid* CV-11) prior to her commissioning on 16 August, 1943 at the Norfolk Navy Yard. (Plank owner) Zollo was the orderly for two captains, one admiral and for an executive officer while in dry-dock when no captain was aboard. He remained on the carrier until she sailed in mid-February 1945.

He was then transferred to the Ammunition Depot at Hingham, Massachusetts where he was discharged on 9 October, 1945. Upon discharge from the Corps, he went back to the boats and received a license and was 2nd Captain on the tow boat. After the war he completed three years of mechanical drafting at Massachusetts Trade Schools.

Anthony Zollo's brother Carmen Zollo.

Zollo met his wife Jennie in 1946 and married in 1948. He left the tow boats in 1952. He became an inspector at the local Alles-Chambers Mfg. Co. (circuit breakers) and after one year became a member of Service and Erection for Alles. (Installed four large air blast brakers at the Aluminum Co. of America Power House in Tennessee) He left Alles-Chambers in 1956 and purchased a garage near his home. Besides selling gasoline, he repaired cars and trucks and had a towing business. He towed heavy equipment long distances - 18 wheelers - Boston to New York. He continued in this business for 31 years. His wife

Anthony Zollo with Bill, Rose, Joe, and Wess at Melody Lane in Frisco 22 January 1944. Anthony Zollo second from right.

became ill and he sold everything to take care of her. She passed away on 8 October, 1988. He has been retired since selling out on 28 February, 1988.

Zollo has a daughter and two grandsons, ages 5 and 7. He also has a son who is a police captain in the Dedham, Massachusetts Auxiliary Police, a Deputy Sheriff and a U.S. Constabul. Both children attended school at Massachusetts State and Northeastern Colleges.

He is a member of the U.S. Naval Institute and the ANA (coin collecting). He is an avid gardener of both flowers and vegetables. He likes to draft, create charts, draw and write. He now lives with his son. He also is a fan of antique cars, owning a 1933 Buick and several Cadillacs.

Zollo is a life member of the *Intrepid's* Former Crew Members Association and visits the ship often. He has made a video tape of the *Intrepid's* travels and history and two other tapes were made for foreign television. He made one tape in his home and the other on the *Intrepid's* flight deck in New York. He was also interviewed by the Discovery Channel in New York on board the *Intrepid.* He stays in contact with many of the original crew of the *Intrepid* and visits them often. The USS *Intrepid's* 50th Anniversary Celebration was held on board the *Intrepid* in New York on 14 August, 1993.

Anthony Zollo with wife Jennie and daughter Brennan on her wedding day in 1984. Due to the nature of his son's police work, he cannot be pictured.

INTRODUCTION

It is a pleasure and an honor for me to write this story on the "Fast Carriers", which includes my ship, the Carrier U.S.S. *Intrepid*, CV-11, CVA-11, CVS-11.

Their performance was unequaled in the history of Naval Warfare. They did the impossible, they roamed the seas at will. They hit the enemy no matter where his force was or how strong his Naval might. Day after day, strike after strike went after the enemy ships and their bases.

This is the story of the Carrier U.S.S. *Intrepid*, CV-11, sister ship to the *Yorktown* and *Essex* before her. When her fitting out and shakedown were completed, *Intrepid* went to war. Every day her men trained, and her fire fighting and damage control units exercised. This training proved invaluable when *Intrepid* burned badly on several occasions. When the ship was under attack, her casualties were always half the number of other ships under attack or afire. Her air groups were the best, no one can dispute the records of Air Groups 6, 18, or 10. Her record of enemy shoot downs and damage to enemy installations and the Japanese Mobile Fleet is unsurpassed by any other U.S. Ship.

I have read and researched many stories pertaining to the Pacific Sea Battles. Many authors refer to the *Intrepid* as the "Other Carrier" or the "Unluckiest Ship". These sources were not fair in their reporting. Because the name "*Intrepid*" was not spoken of and at times completely ignored, I write this story of her travels, her actions, and the bravery of her crew.

Because *Intrepid* is about to celebrate her Fiftieth Anniversary, thirty-one of them with the fleets, this story is somewhat shortened with regard to her actions, in order to honor the ship in her post war travels. Book II will be a daily report of her travels, actions, and participation in the Pacific War.

Having kept a diary while I was a member of her Marine Detachment, from Commissioning into 1945, I logged her everyday travels and duties with the Fleet. No one can deny that *Intrepid*'s record is second to none of the other "Fast Carriers", or that her crew and air groups did not do their best. *Intrepid* saw it all and her crews and air groups performed heroically, especially in actions at the Marshalls, Truk (where she was torpedoed), the Palau Group, Philippines, Luzon, Formosa, Okinawa, the Japan Raids, and the Wake Island Raids (the last strike as the war ended).

Intrepid's nickname of "Unluckiest Ship" started when she hit the Channel wall in a turn on the way to the locks and Balboa, at Panama. A court of inquiry later cleared the ship, its C.O., and the pilot on board of fault. The three hundred foot wide channel was much too small for ships of *Intrepid*'s size. In contrast, nothing was ever said about the *Essex* running aground in Roosevelt Roads in January 1967, where she stayed hard aground for two days. The *Essex* was never called "Unlucky" or other names, as was *Intrepid* early in her career.

The *Intrepid*'s record was equal to, and in some cases surpassed that of the other "Fast Carriers". When orders were given to abandon ship on 25 November 1944, all hands stayed aboard to fight the fires and save the ship. Not one man went over the side to leave the ship that day. It is with much pride and honor that I tell this story of the Carrier *Intrepid*, fifty years later. May she ride the tide at her berth in New York for a long time. Well done *Intrepid*! I am honored to have served her during those hectic days and to now write of her exploits, on this her fiftieth birthday. May she have many more!

Come to and be welcomed aboard the *Intrepid* Sea, Air, and Space Museum, at New York.

The Mighty I

Six days before Pearl Harbor it was begun
And seventeen months later it was done
That the Mighty I was made... and Intrepid was her name
A ship that would sail the seas to fame.

But first to test her guns and crew, and sail the Canal thru
Then to battle for the untried crew
And during '43-'45, her planes downed six hundred of the enemies from the sky
The most hit carriers of that war... the ship that would not die.

And now the Intrepid earned her rest
But in the '50s faced the next test
Assigned to NATO she began her tour
And sailed the Atlantic and Mediterranean Sea lanes galore
And the NASA Missions of '62 and '65
To show that Americas Space Program is active and aware.

And during the Cuban Missile Crisis, the Intrepid was there
To counter the threat of the Russian Bear
And during the Vietnam Affair
The Intrepid did three tours of duty there.

And now her sailing days are done
Her crews and air groups are gone
Her ship's engines no more roar
And her planes fly no more
But a legend and legacy did she leave...

Thirty-one years of faithful duty did she serve
Toughness, Valor and Pride bespeak her name
The ship that sailed to glory and fame
The Mighty I was she
The Mighty I was she.

Eli Kaplan

A Chronology of World War II in the Pacific

1937

July. Japanese and Chinese forces exchange gunfire at the Marco Polo Bridge near Peiping, starting the Sino-Japanese War.
December. The Japanese "Rape of Nanking" in which an estimated 250,000 civilians were killed, shocks the world.

1938

August. Chinese government commander Chiang-Kai-Shek withdraws to Chungking as Japanese occupy most Chinese sea ports.

1939

May - September. Japanese and Soviets do battle at the Mongolian border of Manchuria. Soviet tanks defeat the Japanese.
1 September. Germany invades Poland. World War II begins in Europe.
6 September. U.S. begins patrol of Atlantic coast to 300 miles offshore.

1940

20 April. U.S. increases new ship production.
27 July. Proclaiming the concept of a "Greater East Asia Co-Prosperity Sphere", Japanese aim was to secure raw materials, especially petroleum, from its neighbors.
August. Japan received access to French Indo-China from Vichy, France. Japan then began to harass the Chinese and the Dutch in the Netherlands Indies, the British in Malaya and Burma and also Australians At this time the area east of Indo-China came under the command of General Douglas MacArthur.
2 September. In return for 50 U.S. destroyers to Britain, the British allow the U.S. to build six new bases on British controlled land.
22 September. The Japanese move into French Indo-China. For two years the Japanese have been invading China.
27 September. Japan signs the Tripartite Pact with Germany and with Italy.

1941

10 April. U.S. submarines and destroyers engage German submarines in the North Atlantic. On this day the *U.S.S. Niblack* attacks a German sub.
7 July. U.S. puts troops ashore in Iceland.
4 September. A German U-boat attacks the U.S. destroyer *Greer*.
17 October. Another German U-boat attacks the *Kearny*.
31 October. The destroyer *Reuben James* is lost in the Atlantic, sunk by a German sub.
26 November. United States and Japan talks are dead-locked.
7 December. The Japanese attack Pearl Harbor, Hawaii. The Japanese attack Wake Island as they retire from this attack.
8 December. The Japanese land in the Philippines, on Guam (a U.S. possession) and at Malaya. They bomb and raid Wake Island again. The United States declares war on Japan. The U.S. minesweeper *Penguin* is lost at Guam by aircraft fire.

A Chronology of World War II in the Pacific

10 December. The Japanese at Malaya sink the British ship *Prince of Wales* and the ship *Repulse* by aircraft attack. The British had no air cover and were almost defenseless. Prime Minister Winston Churchill will later say, "*In all the War, I never received a more direct shock.*" The U.S. submarine *Sealion* and the minesweeper *Bittern* are lost at Cavite, Philippine Islands, by enemy aircraft fire.
11 December. Germany and Italy declare war on the U.S.
23 - 24 December. The small Marine force on Wake Island surrenders to the Japanese after a long and heroic defense of the island.

25 December. Hong Kong falls to the Japanese.

1942

2 January. Manila falls to the Japanese
19 January. The U.S. sub S-27 is lost at the Aleutians, and the sub S-36 is lost at Makassar Strait.
15 February. The surrender of Singapore was Britain's single most catastrophic defeat as the Japanese take 130,000 troops captive, many of whom are forced into slave labor in Burma — the infamous "area of death."
18 Feb. The destroyer *Truxtun* is lost off Newfoundland.
19 February. Japanese carrier planes raid Port Darwin, Australia. U.S. destroyer *Peary* is lost at Port Darwin by enemy aircraft attack. The U.S. sub *Shark* is lost in the Molucca Sea.
27 February. In the Java Sea Battle, two U.S. cruisers and three destroyers are sunk. This gives the Japanese control of Java and the Netherlands Indies.
28 February. The Japanese land on Java. The U.S. destroyer *Jacob Jones* is lost to a German torpedo off Cape May, N.J.
1 March. The U.S. light cruiser *Houston* and the destroyers *Pope* and *Edsall* and the tanker *Pecos* are sunk in the Java Sea by enemy surface and aircraft fire. Also lost was the destroyer *Pillsbury* at Bali Strait by surface fire.
2 March. The destroyer *Stewart* is lost at Java when she capsized.
11 March. Gen. MacArthur and his staff leave the Philippines and go to Australia.
12 March. U.S. occupies the Loyalty Islands at New Caledonia. New Zealand sends 157,000 troops into the fight (17,000 of these are Maori warriors, all volunteers). New Zealand supplies British and Americans in that area with food and wool supplies. This area of the South Pacific is now under the command of Admiral William Halsey and his ever growing fleet carrier forces all under command of Fleet Admiral Nimitz.
18 March. Action at New Hebredes, Vila, Espiritu Santo.
March. Japanese occupy Manila and begin strikes and invasion of New Guinea. U.S. carriers strike Lae and Rabaul at New Guinea area.
5 April. Japanese attack Ceylon and sink a British carrier and two cruisers.
8 April. Bataan falls and the Japanese go ashore on Luzon.
18 April. Gen. Doolittle leads 16 B-25s from the carrier *Hornet* to bomb Tokyo and environs. *Hornet* comes to within 620 miles of Japan to launch the planes. *Hornet* is commanded by Capt. Mitscher. Adm. Bill Halsey commands the force.
April. U.S. carrier *Wasp* makes two trips to Malta, transporting British planes.
3 May. U.S. submarine *Perch* is lost at Java Sea by surface craft fire.
4-8 May. Japanese southward drive is stopped at the Battle of the Coral Sea. Japanese lose their carrier *Soho.* U.S. carrier *Lexington* is sunk here and carrier *Yorktown* is badly damaged. The U.S. destroyer *Sims* and the tanker *Neosho* are lost in this battle. The two naval forces never saw each other as they were 180 miles apart. All action was by aircraft. This battle slowed the Japanese in their advance on

Wife of Admiral Hoover, Mrs. John H. Hoover, the sponsor of the christening of the USS Intrepid 26 April, 1943. (National Archives Photo)

The launching of the Intrepid at Newport News 26 April, 1943. (National Archives Photo)

Port Moresby in New Guinea. In this naval battle, the carriers became the commanders of both sea and air battles and so began a new era in naval warfare.
6 May. Corregidor falls to the Japanese after many days of bombardment. Manila Bay is now in Japanese control.
3 - 4 June. Japanese raid Dutch Harbor but are driven off on the 6th.
3 - 6 June. The turning point in the Pacific war begins with the defeat of the Japanese naval forces at Midway. The U.S. had broken the Japanese code and knows every move of the Japanese navy. Forewarned of the Japanese ship movements, the U.S. waits 180 miles north of Midway Island to hit the Japanese naval force headed to Midway. Japanese Fleet Admiral Yamamoto is defeated here. In a five-minute air attack on his carriers, the U.S. planes hit and sink four enemy carriers and hit three cruisers. The Japanese sink the hastily repaired *Yorktown* and the destroyer *Hammann* is lost by enemy torpedoes. Admiral Nimitz' Pacific fleet now commands the north Pacific as well as the south Pacific. As this action takes place at Midway, another Japanese force is headed north to the Aleutian Islands.
7 June. Japanese invade the U.S. owned Aleutian Islands. They take Kiska and Attu.
July. Japanese in the New Guinea area attempt to take Port Moresby and fail.
4 August. Destroyer *Tucker* is lost at Espirito Santo by a mine.
6 August. The battle for Guadalcanal begins and lasts until February of 1943. The U.S. operation here was called "Operation Cartwheel." U.S. forces by-pass Rabaul, isolating it and its Japanese defenders. Some of the battles which took place in that area were: The Battle of Savo Island, 8-9 August; the Battle of Cape Esperance, 11-12 October, and the Guadalcanal battle of 12-15 November. As November ended, the Battle for Tassafaronga (30 November and 1 December) was fought.
7 August U.S. lands troops in the Solomon Islands and Tulagi. Much night action takes place in this area. The Japanese make many attempts to land troops, but U.S. forces beat them off in several battles. The battle at Tanamboga at Guadalcanal takes place as does the Battle at Florida Island. There is action at Henderson Field.
9 August. The Battle at Savo Island takes place; the enemy is victorious. Three U.S. cruisers, *Astoria, Quincy* and *Vincennes,* and an Australian cruiser are all lost by surface craft fire. The destroyer *Jarvis* is lost at Guadalcanal by enemy aircraft attack.
23 August. The Battle of the Eastern Solomons takes place. The Japanese lose the carrier *Ryujo.*
15 September. U.S. carrier *Wasp* and the destroyer *O'Brien* are lost south of San Cristobal Island by enemy torpedoes and the U.S. battleship *North Dakota* is damaged as Japanese submarines come into the area.
Oct. As the month begins, Henderson Field is in U.S. hands.
11 - 12 October. The Battle of Cape Esperance. The U.S. is victorious. Action at Munda and the Solomons.
12 October. The night battle at New Georgia. Destroyer *Duncan* is lost at Savo Island by enemy surface fire.
13 - 14 October. Japanese bomb American forces on Guadalcanal.
15 October. Japanese land on the north coast of Guadalcanal. Destroyer *Meredith* is lost at San Cristobal Island by enemy aircraft attack.
26 October. The battle of Santa Cruz. The U.S. carrier *Hornet* is lost as is the destroyer *Porter.* The *Enterprise* is now the only operational U.S. carrier left in the Pacific.
13 November. The U.S. light cruiser *Atlanta* is lost at San Cristobal Island by enemy surface fire. Also, the *Juneau* is lost due to enemy submarine torpedoes.
13 - 15 November. Battle of Tassasfaronga (Lunga Point). Outcome favors U.S. on balance. Japanese

A Chronology of World War II in the Pacific

lose battleships *Hiyei* and *Kirishima*, one cruiser, and three destroyers (*Akatsuki, Yudachi, Ayanami*). U.S. loses two cruisers and seven destroyers (*Barton, Cushing, Laffey, Monssen, Benham, Preston* and *Walke*).
30 November. U.S. heavy cruiser *Northampton* is lost at Savo Island to enemy surface craft fire.
9 December. Actions at Munda and New Georgia in the Solomons area.
10 December. Action at Gona, Japanese held since 22 July, 1942.

1943

January. Japanese in New Guinea fail to take Port Moresby by the Kokoda Trail area. They withdraw to Buna and Gona and are met by Australian forces and U.S. troops and are defeated, ending the threat to Australia.
2 January. Action at Buna, Japanese held since 22 July, 1942.
5 January. Action at Monda and New Georgia.
10 January. U.S. sub *Argonaut* is sunk at New Britain by enemy surface fire.
30 January. Action at Rendova - Rennell Islands. U.S. heavy cruiser *Chicago* is lost at Rennell Island by enemy aircraft attack.
1 February. Destroyer *De Haven* is lost at Savo Island by enemy aircraft attack.
9 February. Final actions at Guadalcanal, Japanese held since 7 August 1942.
16 February. Japanese evacuate Guadalcanal.
February. U.S. subs *Amberjack* and *Grampus* are overdue at New Britain Island and presumed lost.
2 March. Bismarck Sea Battle. A large Japanese convoy headed to New Georgia is destroyed by U.S. and Australian aircraft.
26 March. A naval action in the north Pacific at the Komandorski Islands takes place.
7 April. Destroyer *Aaron Ward* is lost at Savo Island by enemy aircraft attack.
30 May. U.S. retakes Attu in the Aleutian islands
8 June. Japanese battleship *Mutsu* blows up.
30 June. U.S. lands on Rendova and on New Georgia.
2 July. U.S. lands on Munda
5 - 6 July. Battle of Kula Gulf. Japanese are victorious. Destroyer *Strong* is lost at Kula Gulf by submarine torpedoes. U.S. light cruiser *Helena* is lost at Kula Gulf by enemy surface fire.
13 July. Destroyer *Gwen* is lost at Vella Lavaella by surface craft fire.
6 August. Battle of Vella Gulf. Japanese lose two destroyers, *Hagikaze* and *Kawakaze*. Japanese tried to reinforce their troops at Kula Gulf and Vella but fail.
15 August. U.S. bombs and invades Kiska. Japanese had deserted the island.
22 August. Destroyer *Blue* is lost at Savo Island by enemy surface fire.
11 September. U.S. invades and takes Salamaua.
16 September. U.S. takes Lae. Action at Papua.
2 October. U.S. moves on Finschafen and takes the area.
3 October. Destroyer *Henley* is lost at Huon Gulf by enemy torpedoes.
6 - 7 October. Battle of Vella Lavella. Japanese are victorious.
6 October. Destroyer *Chevalier* is lost at Vella Lavella by surface fire.
7 October. U.S. invades and takes New Georgia Island.
1 November. U.S. lands on Bougainville and defeats Japanese there.
2 November. Battle of Empress Augusta Bay. U.S. is victorious.
21 November. U.S. troops land in Gilbert Islands. Makin and Tarawa bombarded and invaded.

Admiral Durgin inspects the crew 2 October, 1943. (National Archives Photo)

The 20 mm guns at rest, while the Intrepid heads for the Marshalls. The battleships South Dakota, Alabama and North Carolina can be seen from the #10 gun battery to port. *(National Archives Photo)*

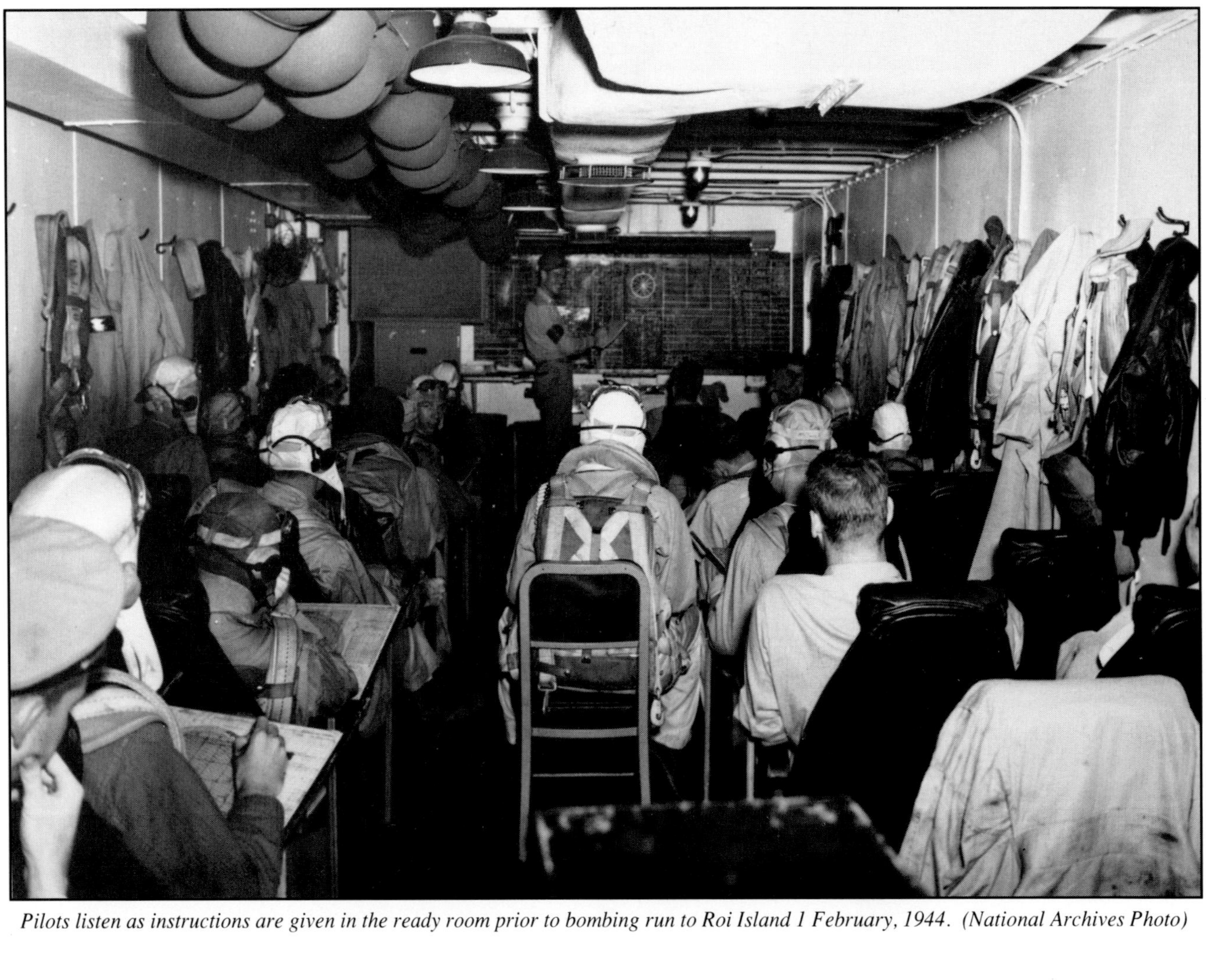

Pilots listen as instructions are given in the ready room prior to bombing run to Roi Island 1 February, 1944. *(National Archives Photo)*

24 November. U.S. escort carrier *Liscome Bay* is lost at Tarawa by enemy submarine torpedo.
29 November. Destroyer *Perkins* is lost at New Guinea in a collision.
15 December. Cape Gloucester and New Britain actions.
26 December. Destroyer *Brownson* is lost at New Britain by enemy aircraft attack.

1944

2 January. Action at Saidor.
31 January - 4 February. Fleet carrier forces bomb and raid the Marshall Islands. Kwajalein, Roi, Namur, Engebe and Eniwetok are blasted then invaded. Kwajalein, Roi and Namur are secured first, then Engebe and Eniwetok.
16 - 18 February. Carrier Task Force moves west to the Caroline Islands. Truk, Hall, Losap, Pulap, and Satawan Islands are all bombarded
17 February. Carrier *Intrepid* is torpedoed at Truk Island at eleven minutes after midnight of 16 February. The carrier is forced stateside for repairs.
18 February. U.S. troops go ashore at Engebe and Eniwetok.
20 February. Engebe Island is secured.
21 February. Eniwetok Island is secured.
9 March. More action at Vella Lavella.
29 - 31 March. Fleet heads west to bombard Palau group.
Spring, Summer and Fall of 1944. The fleet carrier forces move north by west to a position off the Philippines. Raids are carried out over Davao on Mindanao. The fleet then pulls back west of Palau. The fleet then heads west by north to a point off Leyte and Samar. Bombing raids are made there and then the fleet pulls back again to west of the Palaus. The fleet then heads due north then west to off Luzon. The fleet strikes Clark Field, Manila and Aparri on northern Luzon and the area at San Bernadino Strait and the Sibuyan Sea. The super battleship *Musashi* is sunk in the Sibuyan Sea. The fleet then moves north to hit Formosa. During these months, the fleet covers all the invasion routes and in June and July covers the invasion of the Mariana Islands.
22 April. Action at Aitape. Marines move from the Solomons to Hollandia, New Guinea.
23 April. U.S. takes Hollandia.
11 May. Action at Wewak.
17 - 18 May. U.S. invades New Guinea at Arara, Wakde and Sarni.
18 May. Action at Manus.
22 May. Action at Wakde.
29 May. U.S. escort carrier *Block Island* is lost off Canary Islands by enemy submarine torpedoes.
15 June. Second attack on Japan by U.S. aircraft. 47 B-29s from China bomb the Imperial Iron and Steel Works at Yawata, Japan. (The first attack on Japan was the Doolittle Raid of 18 April 1942 when 16 U.S. B-25s bombed five Japanese cities).
12 - 15 June. Massive bombing of the Marianas by the U.S. Task Forces. After much bombing of the Marianas on the 15th U.S. troops invade Saipan. U.S. carrier force engages the Japanese naval forces west of the Marianas.
19 June. The Japanese try to stop the Mariana invasions. The Japanese naval forces are destroyed by the U.S. carrier force. This action is later called the "Mariana Turkey Shoot" as many enemy planes were shot down from the Japanese carrier force. The battle goes into the record books as the Battle of the Philippine Sea. The Japanese lost three carriers (*Taiho, Shokaku* and the *Hiyo).* Lost also was over 90% of their carrier planes with their first line pilots and crews.

A Chronology of World War II in the Pacific

9 July. Saipan is secured.
20 July. Tinian and Garapan are bombed and on 24 July, invaded.
29 July. Guam receives 17 straight days of bombing, then invaded.
30 July. A landing is made on the Vogelkop Peninsula in western New Guinea and then we attack Halmahara, Molukka Passage and Makassar Strait. We hit Mindanao and an enemy ship convoy off its coast.
31 July. Sansapore is taken.
21 August. Guam is secured in the Marianas. The fleet carrier forces move at will over the Pacific.
8 September. A Japanese convoy off Mindanao is sunk. All 32 ships in the convoy go down as do some 20 heavily loaded barges. The ships were cargo ships. These sinkings are made by the carrier planes of Adm. Bill Halsey's fleet carrier force. The planes were on a bomb run over Mindanao when the convoy was sighted. The strike on Mindanao was made to wipe out the airfields and sea ports. Some 70 enemy planes were smashed on their airfields. Some 90 Japanese ships were hit or sunk on this Mindanao raid. The carrier force ships were not hit and the U.S. losses were only a few planes.
13 September. Some 200 fleet carrier planes go up again. This time it is Cebu. Then strikes are made on Negros and Panay. Cebu harbor is aflame and in ruins.
14 September. The fleet heads east back to the Palaus, 300 miles away, after some 28 days at the Philippines. Peleliu is first bombarded and then the 1st Marine Division goes ashore on 15 September. Peleliu is in flames as the Marines hit the beach. In just two days, the 81st Infantry (Wildcat) Division goes ashore on Morotai and Anguar. General MacArthur lands with the troops on Morotai.
16 September. Morotai is secured.
22 September. Ulithi Island is taken. It becomes a major U.S. anchorage and repair yard.
3 October. The U.S. sub *Seawolf* is mistakenly sunk by the U.S. destroyer *Narwall* north of Morotai. The destroyer *Shelton* is lost at Morotai by enemy submarine torpedoes.
9 October. After leaving the Palau group area the fleet, on its way to Leyte, heads first to Okinawa and bombs it. A half dozen ships in the Harbor are left afire as is much of the Island. The fleet then moves to the Philippines.
10 - 16 October. The U.S. fleet carrier force is at Formosa. The airfields at Takao are bombed.
20 October. The Leyte invasion begins. The U.S. Sixth Army goes ashore as the Third and Seventh Fleets stand by. MacArthur has returned to the Philippines. (At Leyte, the Japanese will lose nearly all of their troops which total some 50,000 defenders. The U.S. Army troop losses were 3,504). The Japanese then begin their kamikaze attacks on the U.S. ships with bomb carrying suicide planes. As the Palau and Morotai campaigns are progressing well, the fleet heads back to the Philippines to carry out more bomb runs before the invasion of Leyte which begins on this day. The Japanese naval forces try to stop the invasion of Leyte, and they are nearly all destroyed by the fleet carrier force. This battle is the Battle of Leyte Gulf. The world's largest battleship, the *Musashi*, is sunk and her sister ship, the *Yamato* is damaged. Two other battleships of the Japanese are lost, also. These were the *Fuso* and the *Yamashiro*. Also sunk were the carriers *Zuikaku*, *Chitose*, *Chiyoda*, and the *Zuliho*. Also lost were 10 Japanese cruisers and nine destroyers. All of these sinkings and this action took place in the Sibuyan Sea, Leyte Gulf, Surigao Strait, areas south and west of Leyte and off the northeast coast of Luzon. The U.S. loses the carrier *Princeton* and the escort carriers *Gambier Bay*, and the *St. Lô*. Also lost were two U.S. destroyers and one destroyer escort. Some of the action took place in Ormac Bay, west of Leyte.
23 - 26 October. The battle at Leyte goes on as the Japanese fleet tries to hinder the operation there and they receive their worst losses in the war.

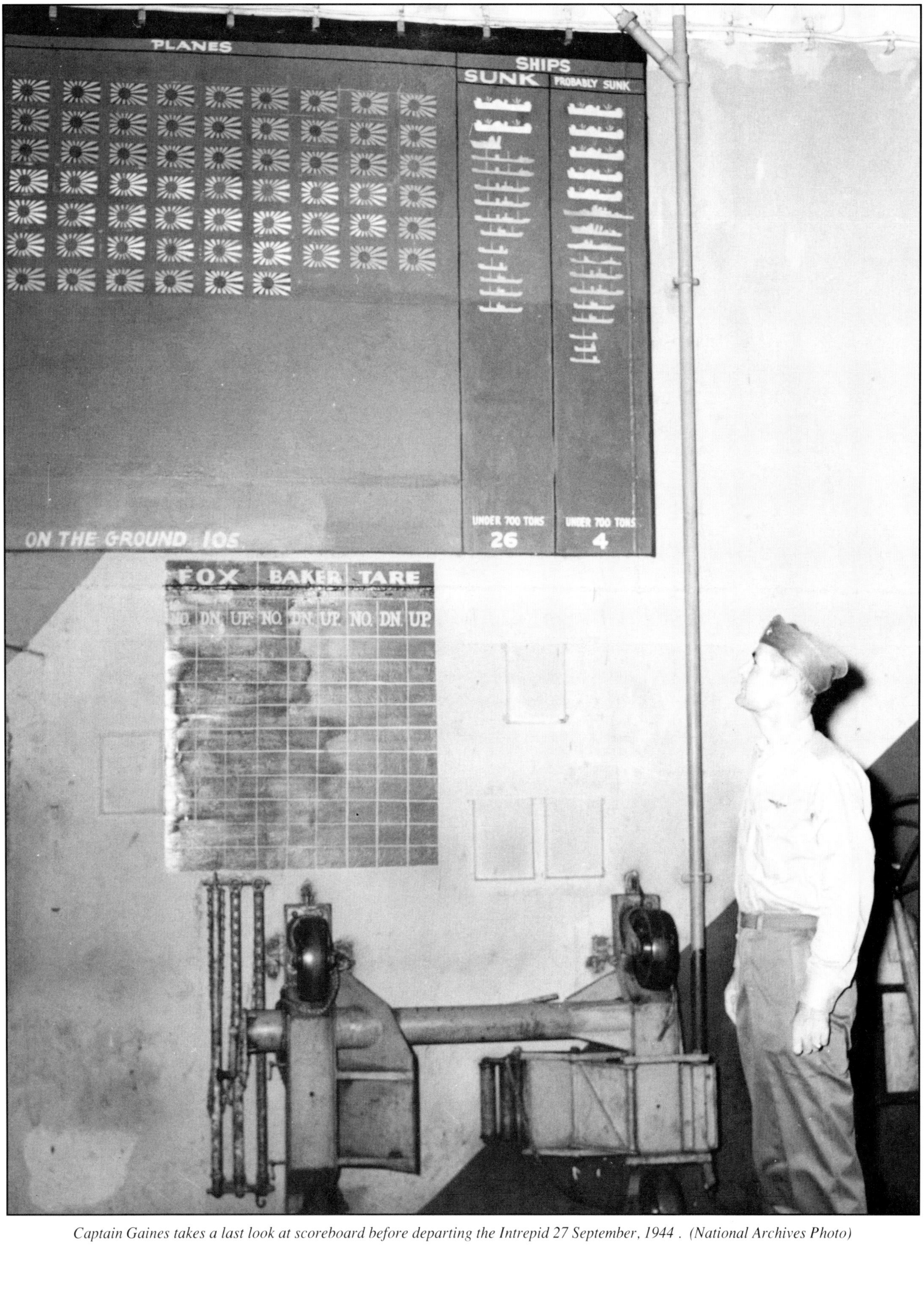

Captain Gaines takes a last look at scoreboard before departing the Intrepid 27 September, 1944 . (National Archives Photo)

Intrepid takes on new planes as seen from escort carrier that transported them. (National Archives Photo)

F6F #28 is moved onto No. 2 elevator to be lifted to flight deck for next strike. (National Archives Photo)

25 October. The U.S. escort carriers *St. Lô* and *Gambier Bay* are lost at the battle off Samar by enemy aircraft attack. The destroyer *Hoel* and the destroyer escort *Samuel B. Roberts* are lost off Samar Island by surface fire.
28 October. The destroyer escort *Eversole* is lost at Mindanao by submarine torpedoes. In October the *Intrepid* is hit twice but repairs her own damage.
30 October. Carriers *Franklin* and *Belleau Wood* are severely damaged.
1 November. The destroyer *Abner Reed* is lost off Leyte by enemy aircraft attack.
19 November. The fleet hits the Manila Bay area. A dozen ships are left sunk or on fire and 118 Japanese planes are destroyed. B-29s bomb Japan. They take off from new bases in the Marianas.
21 November. The U.S. sub *Sealion* torpedoes and sinks the Japanese battleship *Kongo.*
25 November. The *Intrepid* is hit twice by two kamikaze suicide planes. She is almost abandoned, but again she is saved by her damage control and fire fighting crews. She is forced back home again for repairs.
29 November. Japanese carrier *Shinano* , largest ever built and on her first cruise, is sunk by a U.S. submarine.
3 December. The destroyer *Cooper* is lost in Ormac Bay by enemy torpedoes.
7 December. The destroyers *Mahan* and *Reid* are lost in Ormac Bay by enemy aircraft attack.
15 December. MacArthur's force travels 600 miles through enemy areas to the invasion of Mindoro in the Philippines. The invasion goes off without a single man hit as there is no return fire from the shore, but once ashore, the enemy planes come over and bomb the invasion forces on Mindoro.
18 December. The destroyers *Hull, Spence* and *Monaghan* are lost in the Philippine Sea due to weather.
19 December. U.S. sub *Redfish* sinks the Japanese carrier *Unryu.* The fighting at Leyte will go on until 4 January. Total U.S. casualties will be 11, 217, killed, missing and wounded. Japanese losses will be 113,231 men, most killed.

1945

9 January. The U.S. invades Luzon by way of Lingayen Gulf (Japanese held since 27 February, 1942). Admiral Kinkaid's Seventh Fleet and General Walter Krueger's Sixth Army return to Luzon. The troops hit the beach in four different areas.
17 January. The fleet bombs Clark Field and Manila. The field is taken on January 25 and the troops push on to Manila. Another typhoon hits the area in which the fleet is operating. Many ships are damaged. In a similar storm on December 18, three destroyers were lost to the storm. The carriers roll and pitch in the heavy seas. They sometimes list as much as 35 degrees in the big swells. Planes are washed overboard and some carriers even have flight deck damage when the huge waves break over the flight decks.
22 January. After some two and a half years, the Burma Road opens, and so does the Ledo Road. The key road to China had been blocked by the Japanese. The roadway was badly damaged and the U.S. engineers go to work to repair it. The key road extends 1,000 miles to the China border. The road is renamed the Stilwell Road in honor of U.S. General Stilwell.
24 January. The 13th Army Air Force returns to Cavite in the Philippines and bomb it and Manila Bay.
29 January. U.S. forces land in Subic Bay and Grand Island.
30 January. General MacArthur's Rangers hit a prison camp at Cabanatuam and release a personal friend of Gen. MacArthur. He was Colonel A.C. Oliver, held prisoner by the Japanese. Other prison camps are also liberated in the Manila area. The prison at Santo Thomas and Bilibid held 5,000 prisoners of which 4,000 were Americans.

A Chronology of World War II in the Pacific

4 - 11 February. The Yalta Conference, a meeting of the big three, Roosevelt, Churchill and Stalin, lasted for seven days.
5 February. U.S. forces go into Manila. The Japanese have blown all the bridges as they retreated so the U.S. forces take to assault craft and cross the Pasig River into Manila, chasing the retreating Japanese.
6 - 13 February. The 11th Airborne Division attacks Manila from the south while the 37th Infantry hits Manila from the north. There is much street fighting as the Japanese retire.
16 - 17 February. A huge U.S. super force sails to near Tokyo Bay. The fleet carriers launch some 1200 planes against the Tokyo area. The ships here have 15 fleet carriers with cruisers, destroyers, submarines and minesweepers to protect them. All targets bombed were industrial areas.
16 February. We return to Corregidor. We use paratroopers dropped from the air. Our Army Liberator bombers (B-24s) blast the island prior to the invasion. Some 6,000 enemy troops are on Corregidor. The fighting went from cave to cave, and tunnel to tunnel. As the month came to an end, 4,250 Japanese bodies were counted. The paratroopers lost 136 killed, 531 wounded and 8 missing.
19 February. Some 650 miles from Tokyo lies Iwo Jima. Our fleet of 800 ships are off shore and the transports carried 40,000 troops with others held in reserve. The troops go ashore in three areas along the eastern beach after a massive bombing is made on the island, which went on for 72 hours. Fighting was fierce and bitter, but in just a few hours, the Marines have control of the first airfield.
23 February. On this day the flag is raised on top of Mount Suribachi. The official flag raising is held on 14 March.
25 February. The superforts (B-29s) again bomb Tokyo. They bomb 240 city blocks. The planes were over Tokyo for some seven hours and it began to snow before the raid was over.
26 February. U.S. troops mop up Manila.
8 March. A group of Navy nurses are released from the Las Banos prison camp by U.S. troops on Luzon.
9 March. Three hundred B-29s raid Japan. They bomb 16.7 square miles of industrial centers.
10 March. Admiral Russell Berkey's unit of the Seventh Fleet is off Mindanao and for ten days prior to March 10, they pound the island with the big guns. On March 10 the 41st Division of the 8th Army goes ashore at Zamboanga at the southwest tip of Mindanao. The 13th Air Force bombers also hit the area.
12 March. Corregidor is secure.
18 March. "Big Ben," carrier *Franklin* is badly damaged when she takes two direct bomb hits. She returns to the states under her own power for repairs.
26 March. Destroyer *Halligan* is lost at Okinawa when she hits a mine.
27 March. Off Okinawa are the Karama Islands. Prior to the invasion of Okinawa, the U.S. takes these islands and use them to shell Okinawa. The 77th Division goes ashore there and they set up a base from which to hit Okinawa.
1 April. At the Ryukyu Islands (Okinawa) the U.S. goes ashore. All the big guns pound the island prior to invasion. The massive U.S. fleet stands out off Okinawa. After the bombardment, the U.S. 10th Army storms ashore as the ships deploy and circle the island on three sides.
6 April. Destroyers *Bush* and *Calhoun* are lost at Okinawa by air attack.
7 April. The Japanese attempt once more to stop the U.S. invasion of Okinawa. They lose again as their ships are blasted. The super battleship, the *Yamato*, is sunk. She was attacked by 280 carrier planes as were the other ships. She goes down with 2,800 of her crew. (Her sister ship, the *Musashi*, was sunk back in November, 1944 by fleet carrier planes).
12 April. President Roosevelt dies at Warm Springs, GA. Vice President Harry Truman is now President. Destroyer *Mannert* is lost this day at Okinawa by air attack.
16 April. Destroyer *Pringle* is lost at Okinawa by enemy aircraft attack.

Flag at half staff at Ulithi for the death of President F. D. Roosevelt. (National Archives Photo)

Intrepid is hit aft where planes are spotted. Intrepid experiences explosions and fire. (National Archives Photo)

Intrepid has a list as she burns. (National Archives Photo)

17 April. A second landing on Mindanao is made. The Army's 24th Division lands at Paraug and Malaboug on Mindanao.
1 May. Australian forces land on Borneo at Balikpapan.
3 May. U.S. forces are at Davao on Mindanao. (This town fell to the Japanese on 20 December, 1941). The destroyer *Little* is lost at Okinawa by enemy aircraft attack.
4 May. Destroyers *Luce* and *Morrison* are lost at Okinawa by enemy air attack.
8 May. On this day Lord Louis Mountbatten, Supreme Allied Commander, Southeast Asia, declared that the Burma campaign was a success but had some scattered fighting still ongoing.
9 May. Destroyer Escort *Oberrender* is lost at Okinawa by enemy aircraft attack.
11 May. At Okinawa, carrier *Bunker Hill* is hit and she is forced back to the States for repair.
18 May. Destroyer *Longshaw* is lost by grounding at Okinawa. U.S. superforts are again over Japan, this time at Honshu. They bomb the Otake oil refinery.
28 May. Destroyer *Drexler* is lost at Okinawa by enemy aircraft attack.
29 May. The 5th Air Force bombs Formosa.
30 May. At Okinawa, the Stars and Stripes fly over Shuri Castle as mop-up begins.
1 June. B-29s bomb Osaka, Japan. 450 planes raid the second largest city and industrial center of Japan.
10 June. Destroyer *William D. Porter* is lost at Okinawa by enemy aircraft attack. Landings are made at Brunei.
15 June. U.S. Marines continue to mop-up on Okinawa.
16 June. Destroyer *Twiggs* is lost at Okinawa to enemy air attack.
20 June. Cebu is taken in the Philippines. (Japanese held since 19 April, 1942). Negros is secured in the Philippines. In Japanese hands since 20 April, 1942. U.S. forces are at Borneo's west coast, Sarawak, Miri, and Kuching. (Japanese held since December 1941, and January, 1942).
22 June. U.S. forces are at Tarakan on Borneo's east coast. (Japanese held since 11 January, 1942).
23 June. U.S. forces are at Aparri, northern Luzon. (Japanese held since 10 Dec., 1941).
10 July. Balikpapan is taken (Borneo). (Japanese landed here on 24 Jan. 1942).
14 July. Battleship *Massachusetts* is at Honshu, Japan and she shells Kamaiski, a factory town on the shore of Honshu. The carrier fleet is also there and they strike some Japanese naval ships. The battleships *Ise, Hyuga* and *Huruna* are sunk by carrier planes. Also sunk is the carrier *Amagi.*
24 July. Destroyer *Underhill* is lost in the Philippine Sea by enemy submarine torpedoes.
25 July. U.S. forces secure Mindanao. (Japanese held since 9 May, 1942).
28 July. Destroyer *Callaghan* is lost at Okinawa by enemy air attack.
29 July. The heavy cruiser *Indianapolis* is lost off Leyte by enemy submarine torpedoes. *Indianapolis* had delivered parts of the atomic bomb to Mariana B-29 bases.
6 August. The first atomic bomb is dropped on Hiroshima, Japan by orders of President Truman. The U.S. B—29 named "Enola Gay" drops the bomb. At least 100,000 people are killed here. The last Naval action of carrier *Intrepid* in the Pacific is the bombing of Wake Island. The island is set afire and leveled with napalm and high explosive bombs.
8 August. The Soviet Union declares war on Japan and they begin a massive drive into Manchuria and then occupy Sakhalin and the Kurile Islands.
9 August. A second atomic bomb is dropped on Nagasaki, Japan.
15 August. The first radio speech ever by a Japanese sovereign is made. It was made by Emperor Hirohito and he broadcasted to his people the surrender of Japan, unconditionally, to the Allies.
2 September. The formal Japanese surrender takes place on board the U.S. battleship *Missouri* in Tokyo Bay as our massive fleet stands out as far as the eye can see. The time is 0908 hours of September 2, 1945. This author was discharged from the U.S. Marine Corps on October 9, 1945 at the Naval Ammunition Depot at Hingham, Massachusetts.

"FIGHTING I" HISTORICAL CALENDAR

- **December 1, 1941**
  Aircraft carrier Intrepid's construction is begun.
- **April 26, 1943**
  Intrepid is christened and launched.
- **August 16, 1943**
  Intrepid is commissioned into the US Navy.
- **January 29, 1944**
  Intrepid supports the invasion of the Marshall Islands.
- **February 16, 1944**
  The mighty Japanese fortress at Truk is struck by US carrier planes in the most daring and spectacular air raid of the war. Intrepid is damaged by a Japanese counter-attack but is able to limp back to Pearl Harbor for temporary repairs, then on to the West Coast.
- **June 10, 1944**
  Repairs completed, Intrepid departs to report back to the Fast Carrier Task Force for duty.
- **August 29, 1944**
  Intrepid launches air strikes against the Palaus, then moves north to hit the Japanese bases in the Philippines.
- **September 17, 1944**
  Intrepid's group returns to the Palaus to help relieve the beleaguered Marines on Peleliu - one of the bloodiest battles of World War II.
- **October 13, 1944**
  The massive air raids against Formosa destroy 600 Japanese aircraft and 24 ships, thereby setting the stage for the invasion of the Philippines. Intrepid's group receives no damage.
- **October 20, 1944**
  Intrepid supports the landing beaches at Leyte, the Philippines, as General MacArthur makes good his pledge, "I shall return."
- **October 23, 1944**
  US submarines initiate contact with a large Japanese fleet which will evolve into the Largest Naval Battle in History - the Battle for Leyte Gulf.
- **October 24, 1944**
  Intrepid aircraft locate the Japanese main force and initiate the second phase of the epic battle - the Battle of the Sibuyan Sea. Intrepid helps to sink the Japanese super battleship Musashi and damage other warships.That night, the third stage of the battle is fought in the Surigao Strait.
- **October 25, 1944**
  The "Fighting I's" task force races north and destroys the Japanese carrier fleet during the fourth stage of the massive battle.
- **October 26, 1944**
  Intrepid's task group, accompanied by the fastest battleships, arrives off Samar to relieve the escort carriers after the fifth and final stage of the battle. The retreating Japanese fleet is repeatedly bombed and damaged - most never make it out of the Philippines.
- **October 29, 1944**
  Intrepid is slightly damaged by the first kamikaze attack against the Fast Carrier Task Force. Exceptional bravery by a group of black mess cook volunteer gunners saves the ship from serious damage at the cost of their own lives.
- **November 5, 1944**
  The US carriers undertake continuous strikes against Japanese air bases as kamikaze attacks increase in tempo. Over 440 enemy planes are destroyed.
- **November 25, 1944**
  Intrepid is seriously damaged by two simultaneous, bomb-laden, kamikaze hits. Sixty four are killed, 81 wounded. The carrier fights for her life for three hours.
- **November 30, 1944**
  Admiral Halsey inspects Intrepid, orders her back to the West Coast for repairs.
- **February 16, 1945**
  Repairs complete, Intrepid departs for the Pacific.
- **March 18, 1945**
  Off the coast of Japan, Intrepid manages the most incredible escape of the war as a Japanese heavy bomber just misses the carrier and explodes alongside, covering the carrier with flames. The explosion is twice as high as the ship is long. The fires were out in minutes - Intrepid stayed in action.
- **April 6, 1945**
  The Japanese attack the US fleet off

"FIGHTING I" HISTORICAL CALENDAR

Okinawa with 700 aircraft and the remnants of the Imperial Navy. Most are destroyed; Intrepid helps sink the super battleship Yamato.

- **April 16, 1945**
  The US fleet is attacked by 345 Japanese aircraft. Five attack Intrepid simultaneously. All are shot down but one, which crashes into the carrier causing 30 casualties.This is the last time Intrepid will be damaged throughout her career.
- **August 1, 1945**
  Intrepid's last air strike of World War II is against Wake Island.
- **March 22, 1947**
  After returning to the West Coast Intrepid is decommissioned and put in reserve.
- **February 9, 1952**
  Because of the Korean War, Intrepid is recommissioned and scheduled for modernization.
- **June 18, 1954**
  After a major overhaul, Intrepid begins continuous service in the Atlantic and Mediterranean.
- **June 1, 1961**
  Intrepid supports US foreign policy off the coast of the Dominican Republic after Trujuelo's assassination.
- **May 24, 1962**
  As NASA's Prime Recovery Vessel, Intrepid retrieves astronaut Scott Carpenter and his Mercury Capsule.
- **August 10, 1962**
  Intrepid is designated the "outstanding attack carrier" of 1962.
- **October 1962**
  Per presidential order, Intrepid operates as a unit of the Naval Blockage around Cuba, prohibiting the entry of Soviet missiles into that country.
- **March 23, 1965**
  Again performing as NASA Prime Recovery Vessel, Intrepid picks up Gemini's "Molly Brown" with Gus Grissom and John Young. Young will go on to command the first space shuttle flight.
- **April 4, 1966**
  Intrepid in first combat tour off Vietnam.
- **June 1, 1967**
  Intrepid is the last US ship to transit the Suez Canal before the outbreak of the Arab-Israeli War.
- **June 21, 1967**
  The carrier arrives on Yankee Station of North Vietnam for her second combat tour.
- **August 1967**
  The Navy's highest group award, the Navy Unit Commendation, is awarded to Intrepid's air group off Vietnam.
- **July 23, 1968**
  Intrepid arrives back off Vietnam for her third combat cruise. During operations she celebrates her 25th birthday and is selected as "Best ship in the Fleet"
- **November 14, 1969**
  The LEM "Intrepid"" lands on the moon, manned by an all Navy crew.
- **November 1972**
  Intrepid deploys to the Sixth Fleet, having replaced a super carrier, as one of two carriers in the Mediterranean. She's the oldest front line carrier in service.
- **March 1974**
  Intrepid is decommissioned and taken out of service.
- **October 13, 1975**
  An act of Congress designates Intrepid as the Official US Navy and Marine Corps Bicentennial Exposition Vessel at Philadelphia. She celebrates both the nation's and the Navy's 200th birthday.
- **October 13, 1976**
  Intrepid is closed to the public as the Bicentennial draws to a close. Her fate appears to be the scrap heap.
- **March 1978**
  Mr. Zachary Fisher establishes the Intrepid Museum Foundation for the sole purpose of "...saving Intrepid for generations to come..."
- **August 4, 1982**
  The Intrepid Sea Air Space Museum is open to the public.
- **January 25, 1986**
  USS Intrepid is designated a National Historical Landmark by the Secretary of the Interior.
- **July 4, 1986**
  Intrepid participates in the 100th Birthday celebration of the Statue of Liberty.

John Forsyth in his #18 plane comes in for a landing on the Intrepid. (National Archives Photo)

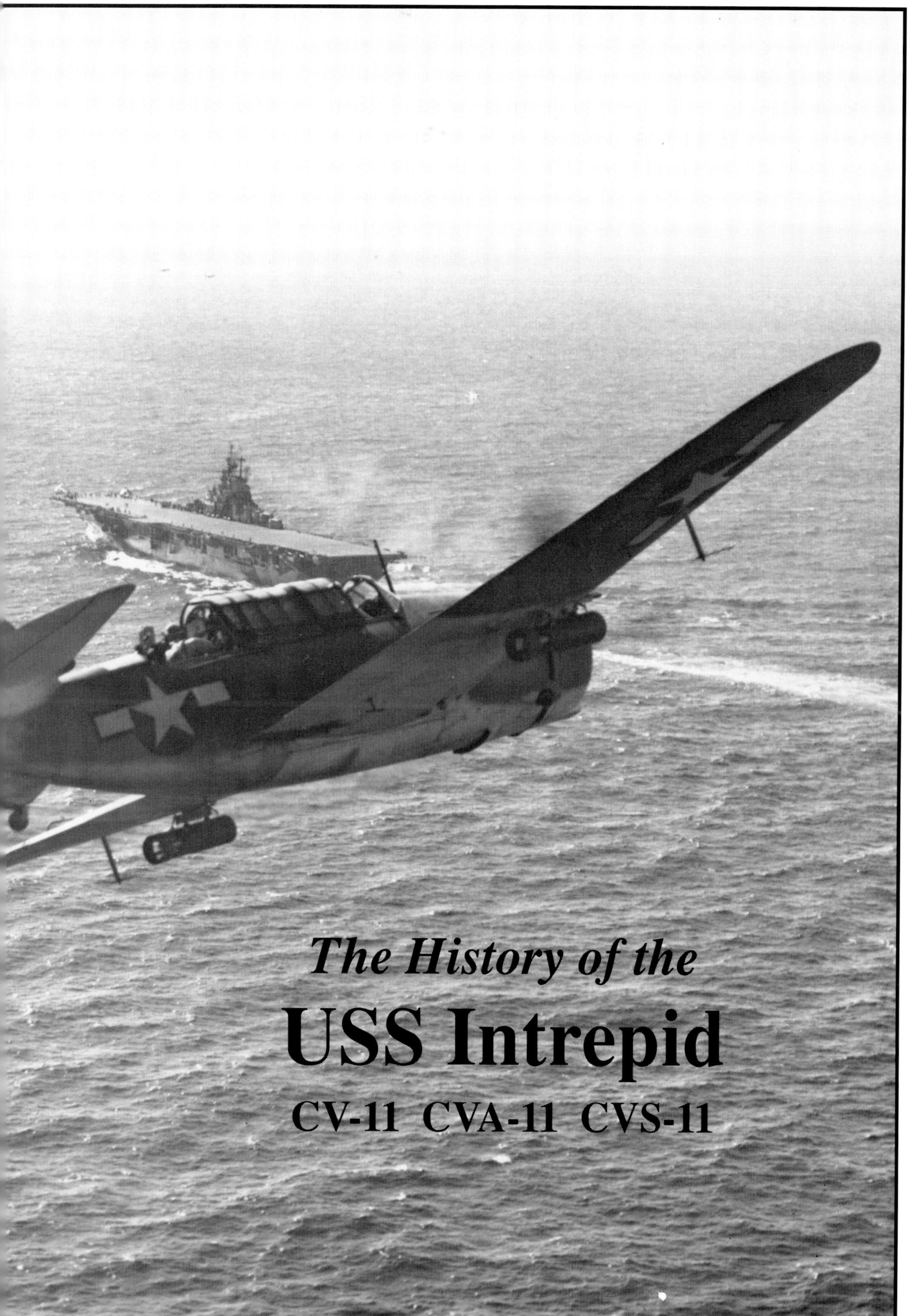

The History of the
USS Intrepid
CV-11 CVA-11 CVS-11

U.S.S. INTREPID

Before the War

Although the United States of America had been aiding Britain with the materials of war since the Spring of 1941 in its war with Germany and Italy, America's declared entry into the conflict did not come until the following December when Japan conducted a carrier plane attack on American naval ships and facilities at Pearl Harbor, Hawaii. War was declared on Japan on 8 December and on Germany and Italy two days later. I noted some interesting parallels between the Pearl Harbor attack at the end of 1941 and a book written in 1925:

British journalist Hector C. Bywater, sixteen years prior to the attack on Pearl Harbor, wrote a story which laid out a plan of attack on Pearl Harbor which the Japanese could carry out and destroy the U.S. fleet at this naval base. So exact was Bywater's plan that the Japanese, sixteen years later, appeared to carry it out to the letter.

The title of Bywater's book was "The Great Pacific War" and in it he spelled out how an attack could be made, the ships involved, the route the attacking force could take — using little traveled sea lanes to the north — and approaching Pearl Harbor and launching its planes southward. He wrote that the Japanese at the same time would invade the U.S. island of Guam, and that they would also invade the Philippines. He predicted that the U.S. would fight back "step by step" (island by island) and carry the war right to Japan's doorstep. Everything that did happen in the Pacific war was nearly a carbon copy of Bywater's story sixteen years before the fact. How could that be?

The attack on Pearl Harbor was planned and executed by Japanese Fleet Admiral Isoroku Yamamoto. As I write this some 50 years later the answer may be that Admiral Yamamoto knew of and read Bywater's book, and that he planned and carried out the attack on Pearl Harbor just as Bywater had written it. Further, succeeding events in the Pacific war between America and Japan were laid out in amazing coincidence and detail in Bywater's book.

Yamamoto was an unwilling servant to his calling; in 1940 he said, "*If I am told to fight regardless of the consequences, I shall run wild for the first six months or a year, but I have utterly no confidence for the second or third year.*"

Franklin Roosevelt was an assistant Secretary of the Navy in the mid 1920's and he and Hector Bywater had a public discussion on whether the Japanese could make war on the United States. Bywater's answer was his book "The Great Pacific War." In this book Bywater also wrote that the Japanese would build the world's largest battleships — and they did: They built the gigantic 68,000 ton Musashi and her sister battlewagon, the equally large Yamato. Was all this a coincidence or could Bywater look into the future?

Among the ten books written by Bywater was one titled "Sea Power in the Pacific", published some four years before "The Great Pacific War." The Japanese purchased many copies of this earlier book and it apparently was distributed among its top navy people. Can we connect Admiral Yamamoto to these books on sea power and war in the Pacific? For it seems that Yamamoto's war plan in the Pacific may have been that of Bywater. Yamamoto was a naval attaché in Washington soon after Bywater's "Pacific War" book was published, and it is known that they attended the same naval conferences on several occasions. (Both men, Bywater and Yamamoto, died before war's end, Bywater in 1940, and Admiral Yamamoto in 1943). One question is: Was Fleet Admiral Yamamoto the great strategist that history says he was or did he at one point read some or all of Bywater's books? These books of Hector Bywater were published between 1922 and 1930 and they tell the story of Pearl Harbor and the Pacific war in detail. Truly an amazing coincidence.

On 1 December 1941, six days before the "Day of Infamy" — the Japanese attack on Pearl Harbor — the keel for a new Essex class aircraft carrier was laid down at the Newport News Shipbuilding and Drydock Company. It would bear the navy identification of CV-11 and would be named *Intrepid*. It would be over two years before she would receive her baptism of war in the Pacific, and it would be a violent baptism.

Pearl Harbor
Sunday, 7 December 1941

The 350 Japanese planes which attacked Pearl Harbor were launched in two waves from six carriers, the *Okagi*, *Kaga*, *Hiryu*, *Soryu*, *Shokaku* and *Zuikaku*. This powerful force was protected by two battleships, two heavy and one light cruisers, and nine destroyers. Also in the area were three submarines of the I type and at least three midget two-man submarines. The task force was fueled by eight oilers.

This strike force had put to sea from the Kurile Islands on 26 November, headed for a point north of Hawaii. The fleet was joined by submarines and they positioned themselves at the sides of the attack force. A few destroyers take the point and are several miles ahead of the attack force. Battleships bring up the rear with cruisers and destroyers at both sides of the centrally placed carriers. These northern waters tend to be rough and they are as the force moves eastward. Some men are washed overboard. There is fog to contend with and some near collisions take place. The seas and the gale winds hamper refueling operations and some of the stored fuel in drums on decks are used. No ships or planes are seen; it appears the attack force is undetected and it will be a surprise attack at Pearl Harbor.

A detached unit of this force was some two dozen submarines. They were to proceed ahead and to detect any U.S. ship movements either coming from or going to Pearl Harbor. They had orders to sink any such ship that was sighted. They were to radio back to the attack force the sighting, course and type of ship that they saw. Five of these submarines carried midget submarines, each with two men inside and their load of torpedoes. These were to head for Pearl Harbor before the aircraft attack. Ahead of the strike force were two destroyers that departed the sub force and headed south at the 180th meridian to bombard Midway Island. This was done under cover of darkness on 7 December.

The force refueled at the 180th meridian on 3 December, and on the night of 6 December the two-man submarines are launched toward Pearl as the fleet itself turned south toward Hawaii. Having done their job, the tankers set a westward course for their home islands. The midget submarines arrive at Pearl Harbor in the early morning hours of 7 December and tried to enter the Harbor under cover of darkness.. At about 0330 hours one midget sub is detected just outside the entrance to Pearl Harbor. The sighting is made by a U.S. minesweeper and it radios the information to one of the night patrol destroyers, the *Ward*, DD 139. The *Ward* made a search of the area but did not locate the midget sub. Some three hours later as daylight breaks, a U.S. flying boat sighted the periscope of a submarine and notified the destroyer *Ward* which is still looking for the reported submarine. At 0640 the *Ward* found then sank the submarine with a pattern of depth charges.

At dawn on 7 December (8 December in Japan), 230 miles from Honolulu, the carriers launched the first wave (a second would soon be launched) of 183 fighters, dive and high altitude bombers and torpedo bombers. This group of planes consisted of 49 of the Kate type, a high-flying bomber which carried one armor-piercing bomb of 1760 pound size; forty other Kate type bombers were loaded with 1760 pound aerial torpedoes. Also in the attack force were 51 Vals, a dive bomber; this plane carried bombs under both wings which totaled 1978 pounds. This group of 140 planes were escorted by 43 Zekes (the Mitsubishi Type 0 fighter) armed with two 20 mm cannons and two 7.7 mm machine guns. The Vals and Kates were equal in fire power, each type with three 7.7 mm machine guns.

The carnage began at 0751 local time. Ships at anchor, airfields, grounded aircraft, buildings—nearly all military installations on Oahu disintegrated under the rain of bombs, torpedoes and machine gun fire. Every field in the area is hit as are buildings, including the hospital areas. Hickam Field is hit by the dive bombers and its planes, which are parked side by side, are destroyed as a few direct bomb hits started a cascade down the line as one exploding plane set off the next. The fields at Kaneoke, Wheeler and Ewa are all hit, bombed and strafed and left in a burning shambles. At Kaneoke Naval Base there was stationed 36 PBY twin engine amphibious flying boats. Three were on patrol at the time of the attacks. Of the 33 on the ground six were damaged and the remaining 27 destroyed. Bellows Field was last to be hit; it received the same destruction as the others. A few Army Air Corps P-40s were able to get into the air and their pilots gave a good account of themselves, but they were no match for the enemy Zekes which controlled the skies. The Army's field at Haleiwa, for some reason, was not hit. Perhaps the Japanese knew it was a training field and had little to offer for targets to bomb. With the parked planes all burning and unflyable, the torpedo bombers could now wreak havoc on the ships in Pearl Harbor with little interference from U.S. planes. The attack now began on the anchored U.S. ships; pass after pass is made on these nearly undefended vessels and they were torpedoed, bombed and machine gunned to twisted masses.

Ships at Pearl Harbor On The Morning Of 7 December, 1941

Battleship *Arizonia*	Moored beside *Vestal*
Battleship *Nevada*	Moored singly—first in the line of ships
Battleship *West Virginia*	Moored to dock
Battleship *Tennessee*	Moored outboard of *West Virginia*
Battleship *Maryland*	Moored to dock
Battleship *Oklahoma*	Moored outboard of *Maryland*
Battleship *California*	Moored at Ford Island inlet
Light Cruiser *Helena*	Moored beside *Ogala*
Light Cruiser *St. Louis*	Moored to dock
Light Cruiser *Honolulu*	Moored beside *St. Louis*
Tanker *Neosho*	Moored at pier
Mine Layer *Oglala*	Moored to dock
Repair ship *Vestal*	Moored outboard of *Arizona*
Battleship *Pennsylvania*	In drydock at Navy Yard
Destroyer *Cassin*	In drydock at Navy Yard
Destroyer *Downes*	In drydock at Navy Yard
Destroyer *Shaw*	Moored alone outside Navy Yard
Hospital ship *Solace*	Moored in deep water (East Lock)
Curtis	Moored at entrance to East Lock
Raleigh	Moored at entrance to East Lock
Target ship *Utah*	Moored at entrance to East Lock

The attacking planes dropped their torpedoes at the battleships which were "rafted" together, two by two, in a line. In well-timed concerted attacks, the torpedoes struck the outboard battleships and dive bombers bombed the inboard ships. The *Solace* was not hit and her small boats picked up the wounded and returned them to the hospital ship for treatment and first aid. Did the Japanese respect the large red crosses painted on the *Solace* or was it a case of so many other targets that she was not attacked? The repair ship *Vestal* was moored outboard of the battleship *Arizona*, and she got underway, leaving the *Arizona's* hull unprotected. The attacking planes made run after run on the *Arizona* and she suffered enormous damage and a great many casualties. A direct bomb hit exploded the *Arizona*'s forward magazines (powder stores) and a series of internal explosions of 500 tons of gunpowder ripped open the big battleship. Hundreds of men were killed instantly and among these were her commanding officer, Captain Franklin Van Valkenburgh, and the commander of Pearl's battleship division, Rear Admiral Isaac C Kidd. The *Arizona* took so many hits that in just nine minutes she sank, trapping hundreds below decks. Later, as the clean up begins, 75 bodies of the *Arizona's* crew are picked up and another 1177 men, both officers and crewmen are killed or missing, entombed in the sunken *Arizona's* hull.

The battleship *Oklahoma* was anchored outboard of the *Maryland* and was one of the first ships to go to General Quarters (battle stations) during this first attack. She took three torpedoes very quickly and assumed a roll-over list and sank. The crew began abandoning the ship over the high side (starboard) as she took two more torpedoes and many bombs from dive bombers which then strafed the escaping crewmen. Her casualties were 429 men killed or missing including 20 of her officers. Below her decks in one water tight area were some men who are trapped; they pounded on the steel plates of the hull and the tapping noises were heard by rescue workers. A hole was burned in the side of the *Oklahoma's* hull and 32 men were rescued. The *Maryland* was protected by the *Oklahoma* beside her and she received only minor damage. Some of the crew of the *Oklahoma* got aboard the *Maryland* and gave the *Maryland's* crew assistance in manning some of the guns. Casualties on the *Maryland* were two officers and two crewmen, all killed.

The *Tennessee*, which had the *West Virginia* outboard of her, received only minor damage but the outboard *West Virginia* took many torpedo and bomb hits and began to list sharply. The port side of the battleship took a beating and most of her guns were out of commission. Cables were fastened to the inboard *Tennessee* to keep the *West Virginia* from rolling over; it worked, and she sank upright in the mud. The *West Virginia* burned fiercely, so much so that the crew had to leave her decks. Many personnel were picked up by a flotilla of small boats now in the water rescuing men. Some of the men of the *West Virginia* managed to get aboard the *Tennessee*, but 103 of her men and two officers were killed in the attack. The *Tennessee* took two bomb hits and sustained some damage and injuries from the burning and exploding *Arizona* just seventy-five feet away.

The minelayer *Ogala* was tied outboard of the light cruiser *Helena*. The *Helena* was struck by a torpedo meant for the *Ogala*, but the torpedo passed under the shallow draft minelayer and hit the *Helena* amidships. The torpedo did more damage to the *Ogala* than to the *Helena*, as her older hull plates burst open. Then a bomb is dropped between the two ships causing further damage to the *Ogala*. A tug put a line on the minelayer and moved the ship ahead of the *Helena*, and the *Ogala*, with her side wide open, rolled over and went down.

Back at the Japanese carriers a second attack wave is launched a few minutes before 0800 hours. This group of planes numbered slightly less than the first wave. This second wave of 167 planes, this time without torpedo bombers, added more death and destruction to the chaos. This second wave flew into heavier anti- aircraft fire as they began the second attack on Pearl Harbor at 0855 hours. Some planes of this second group went for a floating drydock which had the destroyer *U.S.S. Shaw* and a tug boat in it for an overhaul. The *Shaw* took two bombs forward and her bow broke off. A third bomb opened her decks and blew up her fuel tanks spreading fire throughout the ship which then triggered more explosions. The drydock's plates were opened by the explosions and it sank, taking the *Shaw's* separated bow down with it, along with the heavily damaged tug boat. The *Shaw* itself — what was left of it — remained floating. Planes of this second wave also bombed the battleship *Nevada*.

In another drydock was the battleship *Pennsylvania* along with two destroyers, the *Downes* and the *Cassin* and all were hit at the same time as the *Shaw*. Both destroyers were severely damaged and the *Cassin* tipped over on top of the *Downes*. The exploding ammunition on the *Downes* created several explosions and fires which spread to the *Cassin* and both destroyers were total losses. The *Pennsylvania* suffered only one bomb hit and damage and casualties were minimal. The drydock sank as its walls were buckled by the explosions.

Many ships at Pearl at that time were undergoing overhauls and repairs, and many escaped damage in the attacks. One of the scheduled repairs on the *Pennsylvania* was the relining of her big guns (rifles in the Navy). With the *Oklahoma* a sunken wreck and plans made to scrap her, her rifles were removed and reassembled on the *Pennsylvania*. This battleship was the last of the damaged ships to receive attention. Anything of value or use was removed from the *Oklahoma*. She was the oldest of the battleships at Pearl Harbor — a sister ship to the *Nevada*. The *Oklahoma* would be cut up and sold for scrap.

The light cruiser *Honolulu* was another ship which was only lightly damaged. She was tied up at the Navy yard getting some repairs made to her. That work and the damage received by the attack was completed in less than a month and she was back in service. The *Honolulu* was the first of the damaged ships to return to service.

The *California*, hit by two torpedoes, took on a list. She then was hit by two bombs which began explosions in her magazines. All efforts to save the ship failed and she sank in the mud, taking 92 men and six officers down with her. The old *Nevada*, her boilers lighted, was able to get underway and her guns went into action quickly. She took one torpedo forward as she moved into the deep water of the main channel, and several attacking planes dived on her placing at least five bombs forward and wrecking her superstructure and bridge area. Worried that she may sink in the main channel, her skipper turned his burning ship toward Hospital Point where she was beached. There the crew put out the raging fires and counted their dead: 47 enlisted men and three officers. The former battleship *Utah*, converted to a target ship, was hit with two torpedoes and she capsized quickly and went to the bottom, a total loss. Trapped inside her were 58 crewmen who went down with their stricken ship.

There were no U.S. aircraft carriers in port on the morning of the attack. Of the Navy's Pacific aircraft carrier fleet of three, *Saratoga* (CV-3), was on the West Coast of the United States, and *Enterprise* (CV-6) and *Lexington* (CV-2) were far west of Hawaii at 200 and 700 nautical miles, respectively. They escaped the carnage at Pearl, and it is probably fortunate that they did not find and attack the Japanese task force; had they engaged this powerful fleet the *Enterprise* and *Lexington* likely would have endured a severe mauling. Four other fleet carriers, the *U.S.S. Ranger* (CV-4), *Yorktown* (CV-5), *Wasp* (CV-7), *and Hornet* (CV-8) were in the Atlantic. The latter three would quickly be sent into the Pacific.

Arriving over Oahu at the time of the attack were twelve Army B-17s on a long flight from San Francisco. They were unarmed. The Japanese planes over the islands pounced on the Fortresses and they got a double dose from our own anti-aircraft fire from the ground. All the big bombers managed to land safely at various places although several were damaged.

The attack on Pearl Harbor lasted some two hours. The planes returned to their carriers, 29 fewer than took off to hit Pearl Harbor. The Japanese attack force very quickly turned north and left the way they came at a high rate of speed. The submarines in the area did not torpedo a single ship and they pulled away to safer waters. The five two-man midget subs did not hit a ship either. The destroyer *Ward* just prior to the attack, found and sank the first one sighted at the entrance to Pearl. The destroyer *Monaghan* also sank one of them with depth charges after she ran over the sub when the attack on Pearl began. The destroyer *Helm* was also out looking for subs with the *Ward* and the *Monaghan* and she saw one stranded on a reef. The *Helm* fired on the Midget sub but the sub freed herself and got away. Later, this midget sub, which was damaged, was found beached near Bellows Field, Oahu and the one officer on her is taken prisoner. The fate of the last two midget subs is not known for certain, but other ships (the destroyer *Blue* and the light cruiser *St. Louis*) report dropping depth charges on submerged targets.

Most of the Navy Yard docks, buildings and warehouses were not touched, and the oil storage tanks survived intact. Many other ships of various sizes were also untouched as were the other dry-docks. The final toll of ships was three destroyed and 18 damaged; all those damaged — even those sunk — would return to service. 169 aircraft were destroyed and another 130 severely damaged, almost all on the ground. Final American casualty numbers were 2,403 dead and 1,178 wounded. The Japanese lost 29 of the 350 carrier planes that were launched and 64 air crewmen.

All the ships at anchor that morning, except the *Arizona*, the *Oklahoma* and the *Utah*, will come back to hunt the Japanese in the Pacific and take their revenge. Within weeks the *Nevada*, *California* and *West Virginia* will be raised, repaired, outfitted and modernized. The 14 inch guns of the *Oklahoma* will be put onto the *Pennsylvania* later, to replace the worn out rifles that were once on her.

The ships are pumped out with huge suction pumps. Once floated, they are moved to dry-docks and repaired. All battleships are salvaged, repaired and will go back to duty, except the *Arizona* and the *Oklahoma*. They go back modernized and more powerful than before they were damaged.

A tremendous number of men were involved in the salvage operation. Men poured into Pearl Harbor in droves. The repair job on these ships required organized man power: Divers, burners, pump people, electricians, laborers, mechanics, shipwrights, engineers, welders, etc. This powerful group of people, along with naval personnel for direction and leadership, did the impossible in their salvage operations at Pearl Harbor. The divers alone spent 28,336 hours under water and in diving operations to float the battlewagons. The workers at the navy yard also did a masterful job. Battleship row was cleared, rebuilt and enlarged except for the *Arizona* in its resting place which remains as a memorial.

The 29,000 ton *Oklahoma* which settled on her side, was left for months on the bottom before any work was started to right and refloat her. She was eventually brought upright by huge cable and winch arrangements from on shore. The big battleship had one third of her 538 foot long side ripped open by the torpedoes The torpedo holes in her hull had to be closed off before she could be pumped out. The ship was then stripped of all usable material and further repairs were abandoned. The *West Virginia* took six to eight torpedoes in her port side, one tearing off her rudder, and settled upright into the mud. Her side was closed in with wood, made watertight and was then pumped out and floated again. It was some seven months before the *West Virginia* was in drydock for repairs, but work continued on her topside and inside the hull.

Pursuing the Enemy 1942

Japan had already attacked and occupied Manchuria on the Asian mainland in the early 30's and invaded China proper in 1937, controlling much of eastern China before the end of that decade. French Indochina fell to the Imperial armies in 1941, followed during the first half of 1942 by the Philippines, the Netherlands

East Indies, Borneo, Celebes, much of New Guinea, and as far eastward as the Solomon, Gilbert and Marshall Islands. The tiny American garrison at Wake Island, 750 miles northwest of the Marshalls, surrendered on 23 December 1941, and the occupation of three barren islands at the end of Alaska's Aleutian Islands chain marked the pinnacle of Japanese expansion into the Pacific. By mid-1942 virtually all Pacific land and sea areas west of the International Date Line and north of 10 degrees south latitude were firmly in Japanese hands, fortified and strongly defended.

Early in February of 1942 the U.S. Navy had recovered sufficiently from the Pearl Harbor disaster to conduct hit and run raids on Japanese-held island bases in the Marshalls and Gilberts. On April 18 Lt. Col. James H. Doolittle led a small strike force of 16 B-25 Mitchell bombers from the carrier *Hornet* to bomb the heart of Tokyo. The raid did little damage, but it did show that the Japanese home islands were not invulnerable and forced the Japanese to tie up large numbers of defensive aircraft at home. And the daring raid was a great psychological boost to America when there was little to boast about.

Between 4 and 8 May 1942 a Navy force that included the carriers *Lexington* (CV-2) and (the *old) Yorktown* (CV-5, 19,900 tons) and commanded by Rear Admiral Frank J. Fletcher hunted down and mauled a Japanese fleet in the Coral Sea, south of the eastern tip of New Guinea, ending Japanese expansion toward Australia. The Japanese carrier *Shoho* and the *Lexington* were sunk although the latter refused to go down after being hit by two bombs, and two or three air-dropped torpedoes and being gutted by fire. The burned out hulk was finally sent to the bottom by our own torpedoes The enemy carrier *Shokaku* and the *Yorktown* (CV-5) were seriously damaged but both still able to operate. In this engagement, for the first time in naval history, neither fleet sighted the other; only carrier planes were involved in the combat. For the United States, the aircraft carrier came of age in this battle.

Four weeks after the Coral Sea battle Fletcher and Rear Admiral Raymond A. Spruance intercepted a large Japanese fleet heading for Midway Island, about 1,000 miles northwest of Hawaii. They knew the attack was coming for the Japanese naval code had been broken in 1941. The Japanese had hoped to lure the inferior (in numbers) American fleet from Hawaii and win a decisive victory with its enormous force of 100 ships, including four fleet carriers and four light carriers. Japanese carrier planes bombed Midway on 4 June 1942 and American carrier planes from the *Hornet*, the *Enterprise*, and from the hastily (and partially) repaired *Yorktown*, and bombers from Midway immediately attacked the huge Japanese fleet. In the four day battle that followed, from 3 to 6 June 1942, the Navy suffered enormous losses in aircraft (37 of 41 aging Douglas TBD-1 Devastator torpedo bombers were shot out of the sky by enemy fighters in 45 minutes, and carrier-based Grumman F4F Wildcat fighters were running out of fuel and crashing into the sea). Perhaps the most devastating loss was suffered by Torpedo Squadron 8 (VT-8) from the *Hornet:* Early on the morning of 4 June she launched 15 TBD Devastators that found and attacked the Japanese fleet but without fighter cover. All 15 were shot down and only one man of the 30 (Ensign George Gay) survived the carnage to be rescued.

But the tides of fortune turned just as abruptly to favor the American forces. At about 1010 hours on 4 June, 37 Douglas SBD-3 Dauntless dive bombers found and attacked the Japanese fleet carriers *Akagi, Kaga* and *Soryu.* Within five minutes all three had been damaged beyond repair and all would sink with most of their planes before nightfall. But the American fleet had been sighted also and the Japanese attacked about 1400 hours on this same day, 4 June. The "old" *Yorktown* (less than five years from commissioning) took the brunt of the strike, being hit with three bombs and two air dropped torpedoes and was abandoned at 1500 hours. (Old CV-5, now a burned out hulk, refused to sink however until a Japanese submarine put two torpedoes into the smoking ruins three days later). While the *Yorktown* was under attack dive bombers from all three American carriers attacked the Japanese light carrier *Hiryu* (which had launched the planes that were attacking the *Yorktown*) and damaged her so severely she had to be sunk by her own escort the following day. With all their carriers out of action the attacking enemy planes had no base to return to and all were lost.

Japanese expansion eastward had been stopped. The cost to the American side at Midway was high: The loss of nearly 100 carrier-based and 38 Midway-based airplanes; the carrier Yorktown; a destroyer; and 307 dead. The enemy did not fare as well: They lost four carriers, over 320 airplanes, a cruiser (the *Mikuma*), and 3500 dead. The Japanese carrier-born air arm would never fully recover from this disaster, but both combatants were bloody and gasping. Now the rollback of the Empire could begin; and it would begin in the Solomon Islands.

Solomon Islands
July 1942 - January 1943

On 7 August 1942 elements of the First and Second Marine Divisions waded ashore at Guadalcanal in the Solomons with minimal opposition. This state of affairs would not last long, for the Japanese reinforced its units at night with thousands of fresh troops to take the battle to the Marines. The result was six months of fighting that would be recorded as among the bloodiest of the long war in the Pacific. The hotly contested 90 mile by 25 mile island finally was secured by the Marines on 8 February 1943. The *Saratoga*, *Enterprise*, and *Wasp* were offshore in this amphibious assault, providing air cover for the Marines; but this Task Group (61.1) failed to stop the influx of Japanese support troops to the island. On 18 August 19 carrier-based Wildcat fighters and 12 Dauntless dive bombers were flown into Henderson Field, recently captured and renamed by the Marines. They would give the Marines some relief from the constant enemy air attacks. The *Enterprise* was damaged by enemy carrier planes on 24 August and seven days later the *Saratoga* was hit by a Japanese submarine torpedo. Both carriers survived. Also on 24 August American carrier planes found and sank the Japanese light carrier *Ryujo.*

A Japanese submarine caught the carrier *Wasp* (CV-7) southeast of Guadalcanal on 15 September and put two or three torpedoes into her; she had to be sunk by our own forces a few hours later. In the battle of the Santa Cruz Islands on 26 and 27 October the fleet carrier *Shokaku* and the light carrier *Zuiho* were severely damaged but with the loss of the *Hornet* (CV-8) which took five bombs and three aerial torpedoes and was gutted by subsequent fires. She was sunk later by a Japanese warship. This series of disasters left the U.S. Navy without a single, undamaged, operational Fleet carrier in the Pacific. The sea battles raged on and around Guadalcanal, this time battleship against battleship, cruiser against cruiser. The sinking of two enemy battleships in mid-November 1942 left the U.S. Navy in control of the seas around the Solomons.

The battle for Guadalcanal would rage on until the Japanese, suffering 14,000 battle casualties and 9,000 dead from disease, evacuated the island on 9 February 1943. American Marines and Army troops had suffered 5800 casualties in the protracted (6 months) battle for the island. Thus ended Japanese expansion southward and the beginning of the end of their territorial ambitions. The Americans had learned a costly lesson on the ways and means of winning back enemy-occupied lands.....and how to make the aircraft carrier a weapon of terror.

The spring of 1943 saw restricted engagements in the South Pacific. An Allied naval forced defeated a Japanese force in the Battle of the Bismarck Sea between 2 and 5 March, in which the Japanese lost 10 transports and four destroyers while attempting to reinforce their retreating troops along New Guinea's north coast. In June of 1943 Allied ground forces began their systematic advance westward in New Guinea, pushing the Japanese before them. The U.S. Marines and the Army did not rest after the Guadalcanal battle: Russell Island northwest of Guadalcanal fell in February 1943, and the main island of New Georgia northwest of Russell was invaded in July, and Vella Lavella still further northwest on 15 August. Japanese forces had been driven from the Aleutians by the middle of August 1943. In 23 days of bitter fighting from 1 to 24 November, American forces took Bougainville Island, northwest of Guadalcanal, and then shifted operations to the Gilbert Islands where Makin Island fell on 22 November and bloody Tarawa on the 28th. As 1943 came to a close, the American carrier task forces, now beginning to rule the central Pacific, began their air strikes on the Marshall Islands.

Intrepid in dry-dock at Newport News just prior to launching of the ship on 26 April, 1943. (National Archives Photo)

Mrs. Hoover and guests at Intrepid Christening 26 April, 1943. (National Archives Photo)

Mrs. John Hoover christens the Intrepid 26 April, 1943 as the Secretary of Navy looks on. (National Archives Photo)

U.S.S. Intrepid Goes to Sea

On 26 April 1943 the partially completed CV-11, U.S.S. *Intrepid* was launched into her element, the sea. Over the next four months she would be made ready for service. From my diary:

> Mrs. John Howard Hoover, wife of Vice Admiral J. H. Hoover, was the sponsor of the U.S.S. *Intrepid* (CV-11) at Newport News, Virginia, Shipbuilding and Dry Dock Co. on 26 April 1943. Now, on this day, 16 August 1943 the carrier *Intrepid* is commissioned at 1300 hours, at the Norfolk Navy Yard in Portsmouth, Virginia. I am on board with the 7th Marine Detachment and I am now a plank owner of this ship. In command of the *Intrepid* is Captain Thomas L. Sprague; his executive officer (second in command) is Commander R. K. Gains. The Marine skipper is Captain Bruce B Hammond and his second in command is Lieutenant Keagy. Third in command is Lieutenant Nickells; all are in the U.S.M.C. (United States Marine Corps). This is the fourth ship to carry the name *Intrepid*. The name means "fearless and brave." She will prove to be one of the fightingest ships of the fleet. Her first action will come in six months and so will her first damage as she will be torpedoed at Truck Island.
>
> On board the *Intrepid* at her commissioning was the assistant secretary of the Navy for air, Mr. Artemus L. Gates.

A ship commissioning is a carefully orchestrated event. Well before the ceremonies the ship's Executive Officer (Commander R. K. Gains, U.S.N., on the *Intrepid*) issued a memo detailing the uniform of the day, the various committees to be formed, arrangements for meeting and escorting the guests, and all details of refreshments, decorations, transportation, etc. All the various events are timed to the minute — the Invocation, the reading of the commissioning directive, sounding of the Colors, the play-

Intrepid's Commissioning at Norfolk Navy Yard. (National Archives Photo)

All at attention and saluting at the Intrepid Commissioning. (National Archives Photo)

Captain Sprague speaks at the Commissioning of the USS Intrepid. Commander Gains to the right (National Archives Photo)

Following the Commissioning the Marines fall in again on the hanger deck, while the guests mingle. (National Archives Photo)

ing of the National Anthem, reading of the orders by the Commanding Officer, turning over of the ship to the Commanding Officer by the Commandant, setting the Watch, the address (for the Intrepid's commissioning) by the Assistant Secretary of the Navy for Air, the order to "Pipe Down", and the Retreat. On this commissioning day of 16 August 1943, the U.S.S. *Intrepid*'s commissioning ceremonies went like clockwork.

There were three other U.S. Navy ships named "*Intrepid*" before CV-11. The following "History of the U.S.S. *Intrepid*" is here reproduced (in part) as it appeared in the CV-11 commissioning ceremonies brochure:

The first INTREPID is believed to have been built as a bomb ketch in France in 1798, for the Egyptian Expedition of General Bonaparte. It was sold to Tripoli and named MASTICO, and when captured off Tripoli [on December 23, 1803] by the American schooner ENTERPRISE [12 guns], was given the name INTREPID. [The United States was in a Naval war with the Barbary States at the time].

The INTREPID was under the command of Stephen Decatur in his brilliant expedition which resulted in the destruction of the U.S.S. PHILADELPHIA on the night of February 16, 1804. The PHILADELPHIA had grounded [on November 1, 1803] and was in the hands of the enemy [as was her crew]. The purpose of the expedition was to prevent her further use against the United States Naval forces. [Decatur sailed into the harbor, boarded and burned the PHILADELPHIA and retired with only one man wounded]. Later the same year, September 4, 1804, under the command of Lieutenant Somers, the ship [Intrepid] was blown up with all hands in a perilous and fatal attempt to damage enemy shipping in the harbor of Tripoli....

The second INTREPID was built at Boston, commissioned in 1874, brig rigged and iron hull, 170 feet in length, 35 foot beam, 11 foot depth, steam torpedo ram, 438 tons. From August 3 to October 30, 1874, she cruised along the North Atlantic coast trying her torpedoes. From 1875 to 1882 she was in commission at the New York Navy Yard. From 1883 to 1889 she was undergoing repairs and alterations at that Yard; stricken from the Navy list, and sold in 1892.

The third INTREPID was built at Mare Island in 1904 by the U.S. Navy. She was a steel vessel, bark rigged, length 176 feet, beam 45 feet, mean draft 16 feet, tonnage 1800 tons, armament four (4) six-pounders and two (2) one-pounders. She was designed and used as a training vessel. This INTREPID was stationed at Yerba Buena, California. She was sold December 30, 1921...... .

This fourth *Intrepid* was designed as a "short hull" Essex-class carrier. She was the third carrier in the series; *Essex* (CV-9) was first, and *Yorktown* (CV-10) the second. Although 32 Essex-class carriers were authorized only 24 were built, and of these only 17 were finished in time to see action in WW II. The technical data and

A port quarter view of the Intrepid in the Norfolk Navy Yard on September 11, 1943. (National Archives Photo)

dimensions for the carriers in this class are all very similar, with only relatively minor variations. Standard displacement of the *Intrepid* was 27,100 tons; later modifications would increase this to 32,800 tons. Full load displacement was originally 36,380 tons and later increased to 44,700 tons. Overall length at launching was 872 feet which would later be increased to 894 feet 4 inches. Width of the carrier at the waterline originally was 93 feet, 2 inches, increased later to 101 feet. Flight deck width at launching was 147 feet, 6 inches but would later become 191 feet, 11 inches with later installation of an angled flight deck. Standard (unloaded) draft (keel depth) was 23 feet; fully loaded it was 28 feet, 7 inches originally, but increased to 30 feet, 10 inches with later modifications.

The *Intrepid's* motive power was derived from eight Babcock & Wilcox oil-fired steam boilers which generated steam at 850 degrees Fahrenheit at 565 pounds per square inch. This steam turned four Westinghouse geared turbines turning four propeller shafts. Total horsepower generated at the shafts was 150,000. Maximum trial speed with all boilers lighted was 32.7 knots. With over 6,000 tons of fuel oil stored in her tanks, the *Intrepid* could cruise 10,700 miles at 25 knots and nearly 17,000 miles at 15 knots. The Essex-class carriers were designed to carry up to 100 propeller-driven aircraft of, generally, three types: dive bombers, torpedo bombers, and fighters. During the war years *Intrepid* carried at times 110 aircraft. By mid-1944 the on-board Air Group strength was established as 96 planes. This number varied widely in combat engagements, due to combat and operational losses. Any

The first F6F to take off from the Intrepid's flight deck with LT E.L. Flightner as pilot. (National Archives Photo)

The first torpedo bomber to land on Intrepid's flight deck on 16 September 1943. (National Archives Photo)

Admiral Durgin inspects the Intrepid 2 October, 1943. (National Archives Photo)

Gaines, Executive Commander, on the bridge 28 October, 1943. (National Archives Photo)

of these aircraft types could be used in long range search missions and some fighters (usually four) were radar-equipped for night operations. The *Intrepid* carried over 230,000 gallons of high octane gasoline for these aircraft plus spare parts to completely rebuild some 20 aircraft. *Intrepid's* wartime complement, although constantly changing, was approximately 3100 men and 340 officers. She could accommodate, however, 360 officers and well over 3100 men if necessary.

Also changing from time to time was the armament of the Essex-class carriers. *Intrepid's* original 1943 armament consisted of sixteen 5 inch/38 caliber guns in eight twin-mounts, eighty 40 mm and fifty-five 20 mm machine guns or "cannons." (The 5 inch/38 caliber designation measures the bore of that rifle [5 inches] and the length of the barrel [38 times the bore] as 190 inches). In 1944 the number of 40 mm guns was increased to 96 barrels and the 20 mm guns by one more to 56 barrels. The 5/38s could fire a 55 pound projectile a distance of more than 18,000 yards (nine nautical miles) at 15 rounds per minute.

As with all newly commissioned ships in the Navy, the *Intrepid* was made ready for sea trials — her "shakedown" cruise. These trials would take place in Chesapeake Bay, along the Atlantic coast and to Trinidad, British West Indies. Under the command of Captain Thomas Lamson Sprague, U.S.N., as her first skipper, the *Intrepid* departed NOB (Naval Operating Base) Norfolk, Virginia, heading south, on 7 October 1943 with Air Group 8 on board, Commander A. McB. Jackson, Jr. commanding. Commander Jackson had made the first landing on the big carrier's flight deck on 16 September. Five days later found CV-11 at Trinidad where she operated at sea, training her crew and Air Group, until 27 October when she headed back to Hampton Roads, Virginia. Air Group 8 was detached on arrival at Hampton Roads on 1 November. On 25 November the *Intrepid* sailed for Rockland, Maine (without her Air Group) to continue her sea trials. After two days of trial runs in Maine waters, she headed back to Norfolk, arriving there on 30 November. With most of the machinery "bugs" and other problems now corrected, and her crew well along in their training, the mighty carrier was now ready to go to war.

On 1 December Air Group 8 reported back on board and on 3 December *Intrepid* got under way for the Panama Canal. On 8 December she was moored at Colon, on the Atlantic side and the following day entered the canal. The following notes appear in my diary:

The Intrepid heads for Panama and the war. (National Archives Photo)

When the ship was in the locks it was a tight fit, at times a gap of six inches of free area was all the ship had and sometimes it scraped and groaned as it rubbed the sides of the locks...... Later, work parties will go over the side in chairs and touch up the sides of the ship. The captain and the crew were that way — something wrong, fix it; something dirty, clean it up. More than once I have heard remarks on the *Intrepid's* looks, how clean and shipshape she always appeared to be.

The *Intrepid* entered [the locks] at 0700 hours and the pilot took her through what was known as Calibra [Gaillard] Cut (a waterway leading to the locks of Balboa); it was a tight fit for a ship of this size. The waterway was only some 300 feet wide with several turns and bends in the river. Again, I came upon a discrepancy in the story of what took place there in Calibra Cut. I have read some articles which report that *Intrepid* ran aground, that the starboard bow hit the bank on one of the sharper turns. I was always of the opinion that it was our port side of the bow that sustained the damage, the torn hull, the rip in the ships side, and the hole that opened up that was about four feet by four feet. We had some bent plates as well and our bow showed a twist and some metal wrinkles, dents, bends, etc. Engines were put into reverse but our momentum, which was perhaps only eight knots or so, was too great and the ship hit the rocky and steep cliff at the side of the river.

The blame for the accident was not of the doings of the pilot or the ship's officers, a board of inquiry by the Navy agreed. They reported that no one was at fault, technically, and no further inquiry would be held in regards to the accident of the ship hitting the shore. There was some talk that one of the locks had been opened and the surge of that water into the river was a factor in the *Intrepid* hitting the bank.

Intrepid goes forward in locks of the Panama Canal in December of 1943. (National Archives Photo)

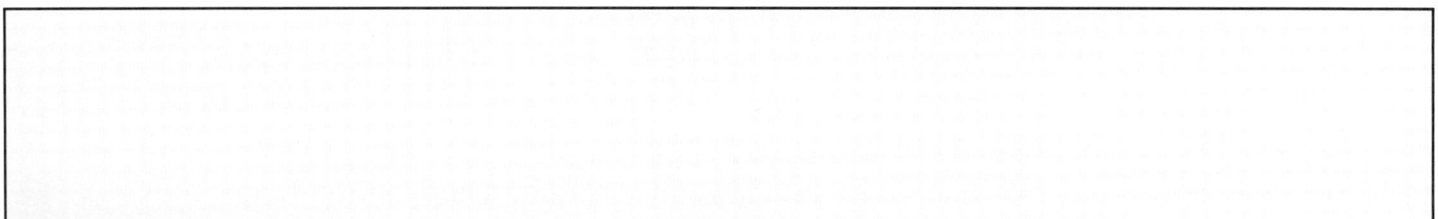

The stern of the Intrepid as she passes through the Panama Canal in December of 1943. (National Archives Photo)

The crew is at ease as the ship enters San Francisco Bay 22 December, 1943 (National Archives Photo)

Midnight Mass on Intrepid at Hunters Point, San Francisco for Christmas of 1944. (National Archives Photo)

Little Mickey Rooney comes on board escorted by two huge Marines 26 December, 1943. (National Archives Photo)

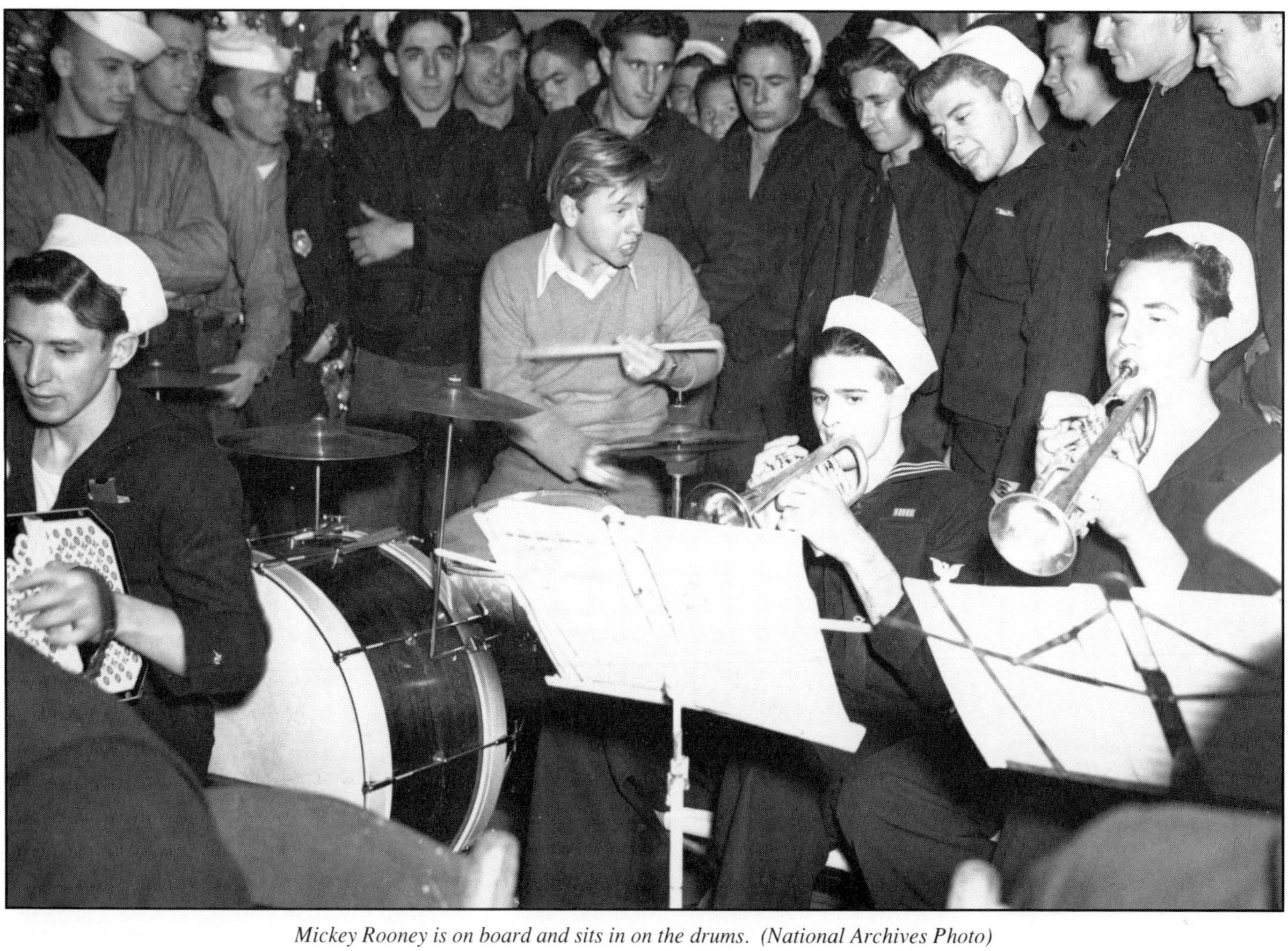

Mickey Rooney is on board and sits in on the drums. (National Archives Photo)

Intrepid leaves San Francisco Bay Bridge behind as she heads for Pearl Harbor. (National Archives Photo)

The *Intrepid* got to Balboa [on the Pacific side] at 1600 hours that same day, and some repairs were made to the bow damage. The hole was covered by a plate welded on. Our stay in Balboa was for five days..... We left Balboa on 14 December for California, arriving at Alameda (Naval Air Station) on 22 December, 1943.

Air Group 8 was disembarked at the Air Station at Alameda (near Oakland) and the following day the big carrier was sailed five miles across San Francisco Bay to Hunters Point for dry docking and permanent repairs to her bow. The repairs would take nearly two weeks. By 5 January of the new year, 1944, the big carrier was again floated and sailed to Alameda to pick up Air Group 8, and the following day left the United States west coast on the passage to Hawaii.

Non-paying passengers on the trip to Pearl. (National Archives Photo)

Marshall Islands Raids

Intrepid arrived at Pearl Harbor on 10 January 1944 after an uneventful four day sail where Air Group 8 disembarked and her new Air Group, 6, reported for duty under the leadership of Commander H. L. Miller, U.S.N. This Air Group went to sea with her new ship for qualification landings on the 12th and 13th, returning to Pearl on the 14th of January. After taking on fuel and provisions, *Intrepid* left Pearl on 16 January 1944, having reported for duty to Commander Task Group 58.2, and headed for war. In company with the fleet carrier Essex (CV-9) and the light carrier *Cabot* (CVL-28) and numerous other ships, *Intrepid* sailed for the Marshall Islands, about 2100 nautical miles from Pearl. T.G. 58.2 would eventually include three battleships, a light cruiser, and seven destroyers. All the major islands in the Solomons and the Gilberts were under Allied control by this time, and the attack on the Japanese Empire would now spread northward to hit Majuro, Maloelap, Mili, Wotje, Kwajalein, and Eniwetok in the Marshalls. The Marshall Islands campaign was code-named "Operation Flintlock." Notes in my diary beginning 16 January read:

> ***16 January 1944.*** *Intrepid* is fueled, provisioned, armed, and ready for war as she departs Pearl Harbor. She is not alone; with her is a great part of Task Force (TF) 58, and other ships will join her on her westerly course to the Marshall Islands. We leave with the carriers Essex and Cabot with us. On 20 January Captain Sprague addressed the *Intrepid's* crew:
>
> *"Now hear this: To all hands of the Intrepid, this is the Captain speaking. The greatest force of carriers, battleships, cruisers and destroyers ever assembled in the world, and the greatest number of auxiliary vessels ever assembled in the Pacific will strike Japanese bases in the Marshall Islands, commencing January 29th. The Intrepid is proud to be in a Task Group of this force, which includes the carriers Essex and Cabot, the battleships North Carolina, South Dakota and the Alabama, the anti-aircraft cruiser San Diego and ten destroyers. Many other powerful task groups will also participate in the total operations and will be in the area.*
>
> *"The Marshall Islands together with the Mariana Islands and the Caroline Islands, were given to Japan under the League of Nations mandate of 1920. It was provided that an annual report be made to the League and that the islands be kept unfortified. No League observers, however, were allowed in the mandated islands, and very few white men went there, and so they gradually became known as the Islands of Mystery. We know all too well that the non-fortification rule was not observed.*

"The Marshalls are a group of coral atolls, and are located at the eastern-most end of the mandated islands, lying about 2,500 miles southeast of Japan and 2,100 miles from Pearl Harbor. These islands form the eastern anchor of the outer screen of bases protecting the Japanese Empire. Although most of the islands are only a few miles in any dimension and have little commercial importance, they are of primary importance both to Japan and the United States as stepping stones for the military movements. Capture of the Marshalls will remove the stony barrier to the Japanese base of Truk, which lies 1,000 miles to the west. Until we have captured the Marshalls, the United States task forces will not be free from air attacks in the move to the western Pacific.

Intrepid recovers planes in January of 1944 underway to the Marshalls. (National Archives Photo)

"The Marshalls differ from the Gilberts to the south, only recently captured by our Naval and Marine forces, in that the Gilberts were only taken by the Japanese from the British in 1942, after war was declared. The Japanese, having had over twenty years longer to strengthen the Marshalls, it is accordingly felt that they may prove a tougher nut to crack. This undoubtedly accounts for the tremendous concentration of force we are throwing against them at this time. This will be the first time we are making

Autographed bombs "To and for the esteemed sons of Hirohito at Kwajalein Atoll from the sons of Uncle Sam are being loaded for Air Group 6 to deliver at the strike in the Marshall Islands. (National Archives Photo)

A personalized bomb loaded for VF-6 to deliver to the Kwajalein Atoll via air express. (National Archives Photo)

a large scale attack at the pre-Pearl Harbor Japanese empire. Landing operations are scheduled, in which our ship (the Intrepid) will play an important role in air support for the invading Marines.

"There are six island land plane bases in the Marshalls, in addition to numerous seaplane facilities. There is an important sub base in the southern portion of the islands. There are excellent fleet anchorages in the many lagoons, protected by the islands and coral reefs of the various atolls. Charts will be posted for all hands to inspect.

"Our targets will be Kwajalein Atoll, the world's largest atoll, being sixty-six miles in a northwest-southeast direction, and has a maximum width of eighteen miles. More than eighty islands and inlets lie along the reef which surrounds a lagoon with an area of 655 square miles, providing more anchorage than any other atoll in the Marshall group. Kwajalein Atoll is the most important Japanese military and naval base in the Marshall Islands. Roi is the target for the Intrepid planes."

20 January 1944. En route to the Marshall Islands Admiral Hill breaks away from the force and heads to Majuro Island to bombard it and other islands prior to the Task Force's arrival there. He is commanding a battleship force which is made up of many of the sunken and damaged battleships of the Pearl Harbor attack two years ago. They were refloated, repaired, modernized and brought up to date with the newest advanced radar and fire power. They now go back to strike the Japanese in their first action since the Pearl Harbor attack. Admiral Hill's battleships and cruisers will bombard the Marshall Islands for days and then they will be joined by the rest of the Force. Admiral Hill has bombarded the islands and now the full force arrives and the carrier planes take off to bomb and strafe the islands prior to invasion. The islands of Roi, Namur, and Kwajalein — all in the Kwajalein Atoll — Wotji and Maloelap are bombed and strafed for almost two days and the bombing continues after the invasions begin. Majuro is the first invaded. Admiral Hill bombarded Majuro very heavily and then the force went ashore on Majuro. Majuro is south and east of Kwajalein Atoll, south of Wotji Island, and north of Makin Island. Hill shelled Maloelap first. This bombardment and these [aerial] strikes hopefully will hold back the Japanese and delay any of their defensive actions against us while the invasions take place. Bombing will continue as the forces land on Majuro. It was reported that Majuro was heavily defended, hence the heavy bombardment. After some three days of bombardment, there is no defense at all by the Japanese forces as the troops walk ashore. A U.S. force bombed this area two years ago; we now invade it.

25 January 1944. We make some additional strikes and bomb runs on Roi and Namur islands. It will take nine days to wrap up these islands and secure them. The landings consisted of some 55,000 [Allied] troops and another 35,000 standing out as reserves.......We destroy all 83 enemy planes on Roi and Namur. We hit any and all installations found there. Much smoke rises from the islands. Ships are refueling at 9 knots. The *Intrepid* is recovering aircraft at 1840 hours. At 1857 hours the Sterret left the formation to investigate a sound contact at our rear. She drops at least three patterns of depth charges. At 1950 hours the Sterret returns to her position in the formation........

26 January 1944. Ships test firing all 5/38, 40 mm and 20 mm guns. All fueling is now completed. Carrier Essex is the flag and the guide. We up speed to 21 knots. General Quarters sounds at 0600 hours. Secured at 0650. Bogies have faded away. Some ships still refueling. The carrier Cabot launches aircraft and the *Intrepid* takes aboard her aircraft at 0945 hours. *Intrepid* recovers 14 aircraft at 1610 hours. *Essex* launches aircraft at 1616 hours. At 1852 hours *Essex* recovers her planes.

27 January 1944. Our position at 1200 hours is Lat. 01°—02.1' N; Long.169°—47.0' E. We crossed the equator at 170°—54' E. General Quarters at 0545 hours; secure at 0646 hours, *Cabot* has flight operations ongoing. She launches four fighters and six TBFs [Avenger torpedo bombers] as cover for the ships [CAP or Combat Air Patrol]. The *Intrepid* is now the guide for the fleet. Admiral Hill bombards Maloelap and Wotji again.

28 January 1944. Sea routines, cruising off the Marshall Islands.

29 January 1944. It is a Saturday and at 1200 hours the ships are at Lat. 08°—54.7' N; Long. 166°—38.2' E. Combat Air Patrol has sighted a submarine periscope at 07 degrees, distance 8,000 yards. The destroyer *Sterret* breaks formation to go and investigate. The *Sterret* reports that the sighting turned out to be a floating spar buoy. The *Sterret* rejoins the Force and takes her assigned position. Land is sighted and General Quarters is sounded at this time. It is 1550 hours. We have several bogies [presumed enemy ships or planes] on the screen [radar] at 105 miles away and closing; now 70 miles distant; course is 195 degrees. Bogies begin to fade away. Secure General Quarters at 1641 hours. One enemy plane on our screen; the destroyer Sterret opens fire on it. He disappears over the horizon. All

The tanker Sabine transports fuel to the Intrepid and the destroyer Lewis Hancock DD-675. (National Archives Photo)

destroyers form anti-submarine positions. We are approaching Roi Island of Kwajalein Atoll in the Marshall Islands. The battleships and cruisers move into positions and bombard Roi and Kwajalein. Roi is first hit and much smoke goes into the air as we hit something that exploded. [*Intrepid* launched its first strike on Roi at 0630 hours] It is 1909 hours and we have a bogie contact at 50 miles. The big guns commence to fire at 1919 hours at the enemy. We are now 20 miles off Roi.

The big guns bombard Kwajalein for three days. Carrier planes bombed and strafed the islands also. Carrier planes with 2,000 pound bombs dropped them on selected enemy defense areas. As the Marines landed on Roi and Namur, it was estimated that seventy-five percent of the Japanese garrison had been killed. At the same time the Army's 7th Division went ashore on Kwajalein Island. Kwajalein was secured in four days, Roi took only 24 hours for it to be secured. Casualties: Kwajalein, 36 dead and missing; Roi and Namur, under 100 killed or missing. During the four days, no ship was damaged, and as a matter of fact no attack was made on the strike force laying off shore. What few planes got off the ground as the invasion began were quickly splashed. My carrier, *Intrepid*, along with the *Yorktown* and others were in the area of the Marshalls from late on the 28th of January to the 6th of February. It was reported that the carriers flew 3,850 sorties and we lost ten fighters, eight torpedo planes but no bombers in the combat strikes. We had arrived undetected and our first wave caught 122 planes on the ground........ Other damages done were two ammo dumps blown up, pill boxes and gun emplacements hit, buildings were hit and three ships set afire. With Kwajalein secure before 4 February came to an end, the Sea Bees were already repairing the airfield and the very next day some land based army planes arrived and within a few days, more planes were on the airfield. These planes went up almost daily and hit all the bypassed islands in the area and other enemy held areas that they could reach. About this time during the war, early in 1944, a phrase was coined, "We gave it a Spruance haircut." This remark referred to the before and after photos of Kwajalein, the second photo showed nothing standing, palms, buildings, etc., all leveled.

The bogie at 1909 hours faded. At 2341 hours we have another bogie at nine miles. This one turns away southward. Secure General Quarters. On this day the invasion begins on Roi, Namur and Kwajalein islands. These islands stretch for about 50 miles in length. Kwajalein is landed on at its most southern point. Roi is located about 45 miles to the north and it is at the widest area of the atoll. There are over 50 small islands in the Marshall chain located in an area at 167 degrees and 9 degrees on the map. Majuro presented no resistance and is secured [on 31 January] and the ships pull into Majuro Lagoon. Roi and Namur are no problem but Kwajalein will put up some resistance. All islands are secured by 4 February 1944.

30 January 1944. The battleship *North Carolina* and the destroyer *Sterret* are bombarding Roi and Kwajalein Atoll. General Quarters sounds early at 0148 hours. Bogies at 210 degrees at 45 miles distant and closing. Bogies turn westward. We secure at 0220 hours. Bombardment continues...... General Quarters again at 0638 hours. We secure at 0713 hours. We have a surface radar contact at 25 miles away on a 285 degree course. These ships are ours. *Intrepid* catapults two planes off flight deck. They will act as our anti-submarine patrol. Roi and Namur are now being bombarded by the battleships *South Dakota* and *Alabama*. After bombarding Nauru [between the Gilberts and the Solomons], Mile, Jaluit and Maloelap in the Marshall Islands, we bypass them and do not land on them. Being cut off, they will fall later on. This massive fleet which is present at the Marshall Islands consists of carriers, cruisers, battleships and destroyers. Also

The airdales spot a fighter on flight deck for take-off on 1 February, 1944. (National Archives Photo)

Airdales spot a F6F which landed from another carrier. (National Archives Photo)

included are the support ships: transports and oilers. There are small craft and large tug boats. A floating dry dock is also in the group which, when in position, will make the area look like a navy yard back home. The Sea Bees [acronym for C.B., Construction Battalion] go to work on the eight airfields here in the Marshalls.

31 January 1944. *Intrepid's* planes strafe Ennuebing Island [next to Roi on Kwajalein Atoll] until a few minutes prior to the troops reaching the beaches. In just one half to one hour the island is secure. This island protects Roi's southwestern flank and controls the North Pass in Kwajalein Lagoon. This allowed the Marine invaders to set up artillery to support the assault on Roi Island.

1 February 1944. At the Marshall Islands, very heavy rain fell on the lst of February and slowed the action some. The *Intrepid* is the guide for the fleet on this day. We have with us the carrier *Cabot*, the battleships *South Carolina, Alabama* and *North Carolina*, the cruiser *San Diego*, and ten destroyers which circle the ships as an anti-submarine screen. The destroyer *Stimble* is ten miles ahead on picket and search duty. We begin fueling operations.

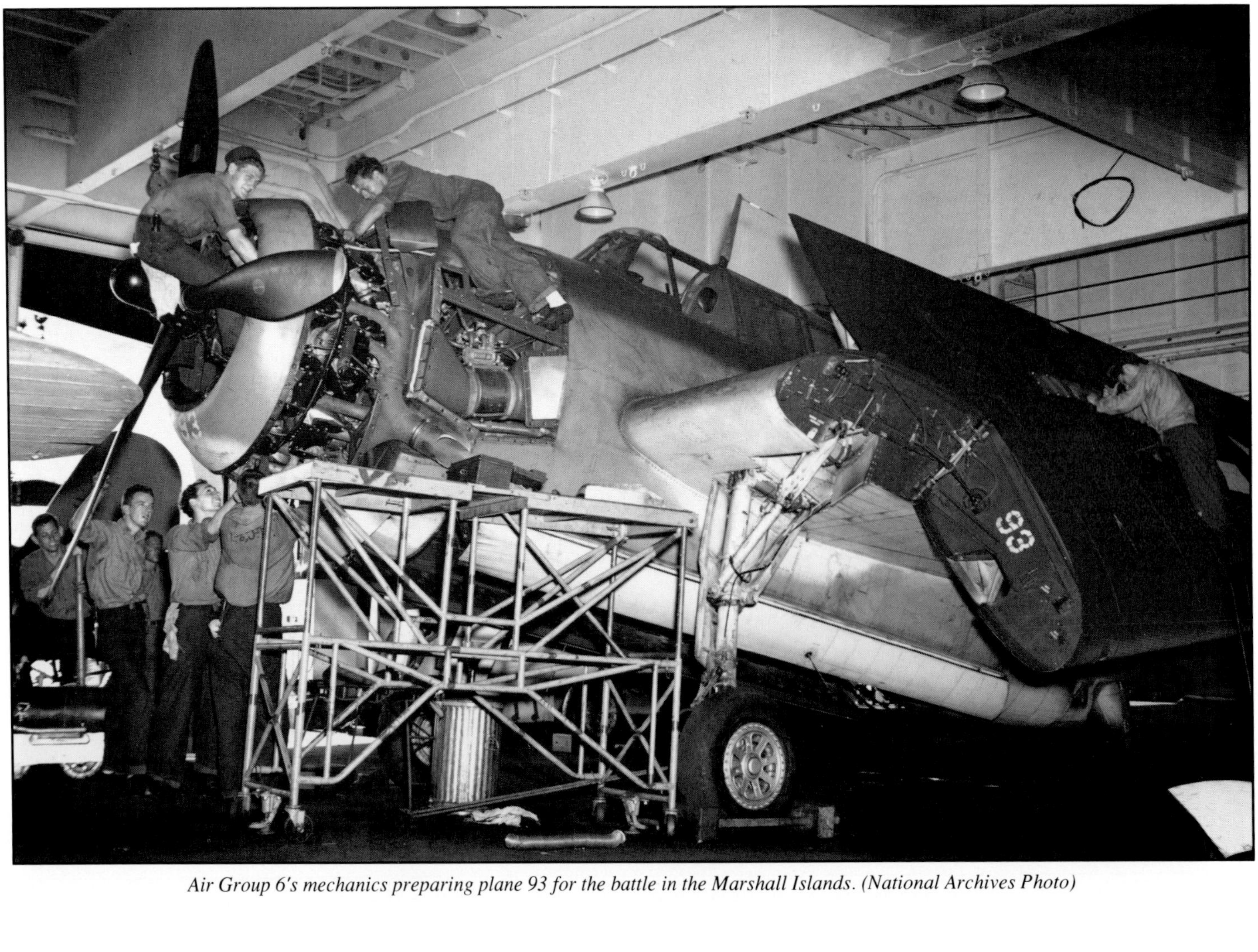

Air Group 6's mechanics preparing plane 93 for the battle in the Marshall Islands. (National Archives Photo)

Of the 528 sorties made through 1 February — 429 were combat sorties — *Intrepid's* Air Group 6 pilots shot down seven enemy aircraft and destroyed numerous aircraft on the ground, losing two fighters and one torpedo bomber to enemy antiaircraft fire. One pilot was lost and one officer was missing in action. In on-board (operational) accidents during these actions, one enlisted man was killed and seven others injured.

Ensign W. A. Davis makes 3000th landing on Intrepid's deck and is congratulated by Landing Signal Officer, Lieutenant R. N. Moot. (National Archives Photo)

3 February 1944. Pilot Ens. W. A. Davies of our VF-6 squadron makes the 3,000th deck landing and is congratulated by the Landing Signal Officer [LSO], Lt. R.N. Moot.

4 February 1944. *Intrepid* is anchored with the fleet at Majuro Lagoon. The water depth at Majuro is 25 fathoms or about 150 feet deep. To anchor we use about 190 to 200 feet of anchor chain, more or less, depending on the spot the ship is over and the water depth at that spot. On this day two specially equipped camera planes (B-24s) are sent to Truk Island to photograph the defensive installations there. They will take pictures at 18,000 feet altitude with special cameras. They will document the island and what is on it.

6 to 9 February 1944. Our sponsor, Mrs. John H. Hoover, presents a good luck plaque to the *Intrepid*. Our dead had been prepared for burial at sea while en route to Majuro. One of our seriously wounded, Everett J. Fure, EM3/C, would die after reaching our anchorage there. Ships refueling en route. The man who died after reaching Majuro was taken to Pergerian Island Cemetery for burial with full military honors. A detachment of Marines are selected to do the honors and I will be part of that detail. Admirals Spruance and Mitscher will take the fleet to Truk Island where we will bombard it for three days. B-24s [from Bougainville Island] photographed the island of Truk on 4 Feb. These pictures are being looked at

Captain Sprague at Majuro in the Marshall Islands. (National Archives Photo)

before we leave to head to Truk. We expect to get underway on the 12th of Feb.

9 February 1944. By the 4th of February the Marshalls were secure but we still had Engebi [Enjebi] and Eniwetok to take. These islands, like the others, are subjected to massive bombing by the carrier planes and bombardment by the big guns. Eniwetok [one of the Eniwetok Atoll islands] is an ugly island. It is a barren, rocky, rough looking place. Engebi [another of the islands of the atoll group] fell on the 18th of February and Eniwetok fell on the 21st of February. In this whole campaign at the Marshall Islands not one U.S. ship is damaged or attacked. The Japanese fleet and their aircraft never interfered with the operations here. In going back over this campaign we make the following conclusions: The islands landed on and secured were Roi, Namur, Kwajalein, Majuro, Engebi and Eniwetok. The islands struck and bypassed were Wotji, Taroa, Mili, Jaluit and Maloelap. Nauru Island was also bypassed. Kwajalein is only 2 and 1/2 miles long and about 2,500 yards at its widest point. At one point, Kwajalein is only 1,000 yards wide. It gave the U.S. the most resistance. The islands are all more or less linked in a long chain which resembles the letter "J." All the islands had some gun emplacements and some pill boxes. A few concrete barricades were also found. Twenty-six islands [of Kwajalein Atoll] of a total of 52 are secured; the rest are bypassed, breaking the chain in several places. As night fell on the first day [of the Kwajalein Island invasion on 31 January 1944] some 12,000 troops were ashore. The 4th Marines went ashore on Roi and Namur islands [on 1 February 1944]. The Army's 7th Division went ashore on Kwajalein Island. [All were secured by 8 February 1944].

On the 17th of February the Marines landed on Eniwetok. Eniwetok lies 300 miles northwest of Kwajalein. The U.S. loss of life was 174 killed and missing and 715 wounded. The Japanese lost 4,650 killed and we took about 50 prisoners. [Eniwetok was secured on 21 February 1944].

Pilots in ready room get final instructions before taking off from the Intrepid on bombing runs to Roi and Kwajalein. (National Archives Photo)

Squadron in ready room prior to taking off to strike the Marshall Islands. (National Archives Photo)

VF6 Squadron on Intrepid 8 February, 1944. (National Archives Photo)

Services are held for Everett J. Fure, EM3/C at Majuro Island. Gunnery Sergeant Crenshaw leads the Marine Honor Guard. First Marine in line is Corporal Page. (National Archives Photo)

The body of E. Fure is taken ashore to the Island of Majuro for burial on 9 February, 1944. (National Archives Photo)

Looking at the Intrepid from the carrier Essex off Caroline Islands at daybreak on 16 February, 1944. Planes are already over Truk Harbor on a bombing run. (National Archives Photo)

Truk Raids

Many of the ships forming the huge Task Force 58 were anchored at Majuro lagoon for about a week, fueling and provisioning for the next operation. Lost planes and personnel were replaced on the carriers, all ships were rearmed with bombs and ammunition, and the force made ready to attack the enemy again. Target: the Truk Island group, about a dozen small islands in a lagoon surrounded by a coral reef about 40 miles across in its longest dimension. The Truk strike was a cover for the amphibious assault on Eniwetok, 300 nautical miles northwest of Kwajalein and 700 nautical miles northeast of Truk, scheduled for 17 February. Truk is about 900 nautical miles west and little south of Kwajalein. It was a major Japanese naval and air base that had to be neutralized. The Fast Carrier Task Force 58 would form up on 14 February at sea, on the way to Truk. It would consist of three Task Groups: TG 58.1 with the carriers *Enterprise* (CV-6) and *Yorktown* (CV-10), and the light carrier *Belleau Wood* (CVL-24); TG 58.2 with the carriers *Essex* (CV-9) and our own *Intrepid* (CV-11) and the light carrier *Cabot* (CVL-28); and TG 58.3 with the carrier *Bunker Hill* (CV-17) and the light carriers *Monterey* (CVL-26) and *Cowpens* (CVL-25). Five CVs and four CVLs, nine in all; together they could launch over 600 planes. In the company of six battleships, ten cruisers (heavy, light, and anti-aircraft types), 29 destroyers plus submarines and auxiliary vessels, Task Force 58 had enormous strike power. Since the fly over of the photo plane on 4 February the Japanese on Truk knew they would soon be targeted. They sent most of their warships westward and out of harm's way. From my diary of 12 February at Majuro:

> ***12 February 1944.*** At Majuro about to leave the lagoon late in the day...... Also leaving are the [cruisers] *Santa Fe, Mobile, Biloxe, San Diego, Oakland, Salt Lake City* and *Wichita*; the battleships *Pennsylvania*, and *South Dakota*; and destroyers *Bronson, Schroder, Abbot, Stack, Cabbot, Martin, Kidd, Izard, Charrette, Conner, Wilson* and *Radford*. With the *Intrepid* are the carriers *Bunker Hill* (CV-17) *Enterprise* (CV-6) and *Yorktown* (CV-10).
>
> ***13 February 1944.*** At sea. Routine. Ships refueling while we head west toward Truk Islands.
>
> ***16 February 1944.*** We arrive undetected at dawn. The Japanese strike on Pearl Harbor on December 7, 1941 with about 260 planes was nothing compared to what they are about to receive from the carrier task force off its shore this day. It is dawn, 16 February 1944 and over three hundred carrier strike planes take off from the carriers to strike Truk Island. The damage done is awesome. The big guns of the battlewagons and

The men who get the planes in the air and back on the deck. (National Archives Photo)

Intrepid's Night Fighter Group AV(N)-101. (National Archives Photo)

cruisers pound the island hard and steady for many hours. Some 30 ships are sunk in the harbor and over 200 planes are destroyed on the ground. Some ships try to leave the harbor and are hit by the big guns in front of them. The carrier planes have also wreaked havoc on the island with their bombs and the strafing. It was said that Truk was not to be taken easily. It ceased to be a threat in two days and we decided to bypass it for now, to be taken at a later date....... 200,000 tons of merchant ships and war ships all sunk. The ships involved at the Truk strike [Task Force 58] were 9 carriers, 6 battleships, 10 cruisers and 29 destroyers. The Japanese lost 8 warships and auxiliaries and at least 22 merchant ships; 250 planes were destroyed or damaged in the air or on the ground, and the islands left afire.

Our Air Group (6) that day had a good day over the Truk Island areas. Shore installations and many ships were hit. Intrepid planes made many passes over the area as did the planes of the other carriers that were with us at the Truk raids. One of Intrepid's targets was the seaplane base at Dublon Island and our planes wasted no time in destroying it completely. On that raid, my notes state that we had sent off on the first strike 12 F6F fighters with the Air Group Commander H.L. Miller in the lead. Also off were nine TBFs [torpedo bombers] loaded with incendiaries and fragmentation bomb clusters. Also off on the first strike were the SBD's [dive bombers]; they carried 500 pound bombs and twelve took off.

The fighters were the cover as the Avengers [TBFs] hit the seaplane ramp area knocking out the seaplanes there. The dive bombers went for the ships which were in the harbor; at least 14 miscellaneous sized ships at anchor there. Some Japanese planes took to the air to intercept our planes. The fighters around the TBFs immediately shot several of them down. The Japanese planes were of the Val and Oscar type and they were no match for our planes. Ground anti-aircraft emplacements were hit as were buildings, aircraft on the ground and the ships at the dock area, and anchored in the channel.....

The strike, as usual, took off just before dawn [0645 hours] and by 0830 hours much damage had been done to Truk. At the seaplane area and the plane ramps, some 15 Japanese sea planes are shown in the photos, all wrecked and half sunk at their moorings. In the channel, *Essex*'s aircraft are hitting a ship of some 14,000 ton size. It was an ammunition ship and *Essex* planes planted four 500 pound bombs on it and it was well in flames. An Intrepid plane came along and dropped the fifth 500 pound bomb on her and she went sky high with several huge explosions. In the briefing after the planes returned, I managed to get additional information on our strike on Truk. Notes state that one of our SBDs put two 500 pounders on what was listed as an aircraft ferry, and on a troop ship, both left afire with much smoke. An *Intrepid* plane had made a direct hit on the troop ship with a 500 pound bomb. Other carrier planes are hitting other ships at the same time. The first wave now moves up to a higher altitude as the second wave comes over the islands.

The planes of the second wave had a field day also, as they are not bothered by the enemy planes which were taken care of by the first wave on the target. In the second wave the torpedo bombers came in and hit the shipping. The SBDs this time had a 1,000 pound bomb, SAP type (semi-armor piercing). More than one cargo ship received a direct hit with these bombs and they just disintegrated in all directions. A Japanese cruiser was listed as one of the ships in the harbor and it was left dead in the water; then the torpedo bombers put two "fish" into her. As night came on, the planes returned to the carriers. Some night flyers went up after dark on another strike. All during the day, and it was a long day, Combat Air Patrol and Anti-Submarine Patrols cover the ships and screen the whole area.

Intrepid planes flew 193 sorties over the target and our ASPs [Anti-Submarine Patrol] ten to twelve sorties, while our CAPs flew eight sorties. Damage done will not be known until a little later, after the photos are developed, but we do know that as of this time, 16 enemy planes have been shot down and at least 38 others hit on the ground [the final number would be 43 for this day]. As the first day comes to an end and everyone settles down for the night, I do some writing and at midnight, I decide to get some sleep.

As noted in my diary, just before dawn at 0645 hours at their launch point 90 miles from Truk, the carriers flew off 72 F6F Hellcat fighters to engage enemy aircraft that may be airborne and to strafe those caught on the ground. The Hellcat pilots shot down 30 enemy aircraft on this first strike and destroyed numerous others on the ground. Then the carriers sent in their dive bombers (SB2C Helldivers) and torpedo bombers (TBF Avengers) to attack shipping and shore installations. The attacks lasted through the day and into the night as Truk was shattered and burned. When it was over, the enemy had lost about 30 ships including two cruisers, four destroyers and two submarine chasers. Four other destroyers were damaged severely. *Intrepid* planes were credited with sinking two ships and damaging ten others in the lagoon.

At dawn the next morning (17 February) the carrier attacks resumed but were called off at noon when the attacking aircraft reported there were no targets left worth hitting. In the 'round the clock, 30 hour strike, the Task Force had lost 25 aircraft (eight of these operationally and not due to enemy action) and 29 pilots and air

Truk Harbor on February 16, 1944 as Intrepid planes make run after run on Japanese shipping. Smoke rises from targets already hit ashore. (National Archives Photo)

Photo taken by Intrepid planes shows two ships under attack, one just gone down and others on fire. (National Archives Photo)

Ships in Truk Harbor as night comes. The Intrepid is hours away from being torpedoed here at Truk. The torpedo comes eleven minutes after midnight 16 February, 1944. (National Archives Photo)

A Japanese cruiser is hit by Intrepid planes and is on fire at Truk. (National Archives Photo)

The Intrepid planes have a field day as no A-A fire comes up from the enemy ship's guns. Eight ships are now on fire. (National Archives Photo)

A different view of the Japanese ships hit. (National Archives Photo)

At Truk Island during the 22nd sortie, a Intrepid plane banks on another run on a Japanese ship already suffering from prior hits. (National Archives Photo)

crewmen. *Intrepid's* Air Group 6 lost seven of its aircraft, six pilots and four enlisted men in these strikes. Truk was no longer useful to the enemy; it would wither and die.

But for the *Intrepid* all did not go well. At the end of the first day's strikes the Japanese sent out a half dozen radar-equipped night bombers (twin engine Bettys) to hit the carriers. It would be the enemy's only retaliatory action. All were driven off by Hellcat CAP (Combat Air Patrol) night fighters - except one. My notes for this date show the following:

It is almost midnight of 16 February 1944. I am in my bunk and still awake. I had been bringing my diary up-to-date. For several days I had numerous notes in my pocket and in my trunk at the foot of the bunk. Tomorrow night I may get back to the chart I've started which shows Intrepid's daily position. I have most of the names of the islands also printed on the map. I have the position of the ship for the past four days to add and I might just get on it tomorrow.

It is a minute or so after midnight. I feel like I'm going to sleep....what seemed like a long time was only a minute or two when there was a loud noise and I find myself flat on the cold steel deck. I have on only undershorts and socks. I realize that the ship had hit something or something hit the ship. I grab my pants and my shoes and I start for the ladder to go up to the hanger deck. The hatch is slammed closed just as I reach the ladder. I turn and I am now at a run and go forward toward officers country. I pass several ladders and hatches and they are shut. I continue on and finally find an open hatch and up to the hanger deck I go. As I enter the island to go to my gun, I no longer have my shoes or my pants. I continue on. I get to the gun which is outboard of the island structure. The 20 mm gun is in tub #3, the gun is #1 of five. I grab a full ammo magazine as I pass through the ready room out into the tub. I turn to the left and slam the magazine into the gun and only then do I look out to starboard to see if I can see anything. About the time

I got to the flight deck and as I entered the island, the PA system came alive and General Quarters is sounded. We are told that the ship has been torpedoed and to man your battle stations.

As I look to starboard from tub #3, I see a twin-engine plane perhaps 500 yards off the ship's beam. The left engine of the plane is spitting red fire as though the engine is missing and not running smoothly. The plane, which I believe was a Japanese twin-engine Betty, was only some 60 feet or so above the water. He then turned sharply left and he flew right across the *Intrepid*'s flight deck from starboard to port. I don't believe the plane was higher than 35 feet above the flight deck as he passed over. He disappeared toward our port side and headed, I believe, to the south. Not a shot was fired.

Torpedo us he did. The torpedo hit the *Intrepid* under our stern, starboard side [the torpedo hit 15 feet below the waterline]. The torpedo penetrated into the ship some distance. The hull was ripped open from the keel to perhaps the fourth deck, a huge hole had opened up. The torpedo cost us eleven men killed and seventeen injured. ....this enemy plane was tracked on radar scopes of more than one ship that night. The carrier *Yorktown* (CV-10) sent up a night fighter to intercept this bogey at a time prior to the actual torpedoing. The record shows that the flight went off and was airborne at 2349 hours. One minute later contact was lost on the position of the bogey by the *Yorktown*, but *Intrepid*'s radar scope still showed the blip. It was still headed to *Intrepid* from a course to her rear. The blip on the screen would occasionally disappear as the enemy plane was really low on the water.

Watching all this, Admiral Montgomery gives orders to take evasive action by making an immediate emergency turn to a new course. Task Group 58.2 at a speed of some 25 knots, goes to course 240 degrees. In the next few minutes during the course change, the bogey is lost on the radar screen. It is now 0008 hours of 17 February 1944. The course change happened just seconds prior to the torpedo hitting *Intrepid*'s starboard quarter. Admiral Montgomery orders another immediate turn from 240 degrees to 150 degrees, but it is now too late. The torpedo finds its mark...... The record shows that we were hit by this torpedo when the bogey was only 300 yards astern of us, and that he was only 50 feet above the water. [Official action reports of 17 February 1944 record the torpedo hit at 0011 hours].

What was this guy doing flying around at midnight anyway? [Night attacks by either side were rare events]. Certainly he found the *Intrepid*'s wake long before he saw the ship and so her course perhaps didn't matter.

Chaplain leads prayer for burial at sea 18 February, 1944. (National Archives Photo)

Eight men to be buried at sea following the torpedo hit. (National Archives Photo)

As I landed on the deck [from my bunk] the ship shuddered and shook some, and seemed to vibrate. I recall that there was hardly any confusion. The men ran and moved in all directions, but no confusion. Each guy knew what he had to do and where he had to go, at least this is the picture I got in the Marine's compartment one deck below the hanger deck at the center of the ship.

The records state that the rudder was damaged and locked at a 15 degree left rudder position. When the ship was put in dry dock at Pearl Harbor, I personally took a look at our stern end as the flooded dock was drained of water. What I saw was this huge rudder hard over to the port side and completely disconnected at the top edge, holding only by its skeg at the lower end. The rudder did not appear to be damaged, that I could see......

Later I heard that the ship lost two of the 20 mm guns aft, torn loose by the torpedo and dumped into the sea. The catwalk along the starboard edge of the flight deck aft was also torn loose and part of it dangled and swung back and forth. Two planes on the after edge of the flight deck went into the water after bouncing on the fantail. They were two SBDs. [Nine other Helldivers were damaged by the blast]. My notes state that two of the dead were CPOs (Chief Petty Officer) who had decided to have a cup of coffee before turning in for the night. [Six men were killed in the explosion and five were listed as missing in action and presumed dead; seven men were seriously injured and 37 others sustained minor injuries]. The torpedo hit and damaged that whole area below and above that section [the steering engine ram and motor rooms were ruptured and flooded by the explosion, jamming the rudder whose top mounts were badly damaged. Propellers, shafts and engines sustained no damage, however].

......The ship is acting funny. We seem to be turning. We are almost hit by the carrier *Essex*. We turned in across her bow. There is no collision but it was damn close. Torpedo Defense sounds after we are torpedoed. No one knew that a plane had entered our screen, that is because he was so low on the water that radar did not pick him up. The PA system says that we have no steering; this is the reason that the *Essex* and *Intrepid* almost collided with each other. The ships give us a wide berth and we begin to straighten out but we cannot keep a straight course. Captain Sprague is alternating our four props both forward and reverse at the same time to try to keep the ship

VF6 Ace Lieutenant (jg) Alexander Varciu on Intrepid with nine shoot downs. Varciu left the Intrepid in November of 1944 with a total of 19 shoot downs.

straight. It is now obvious that we have no rudder and we have no control to keep the ship straight. In his report the Captain said, "The ship would weather-cock with the wind. The ship kept turning west towards Tokyo; at the time I didn't want to go to Tokyo."

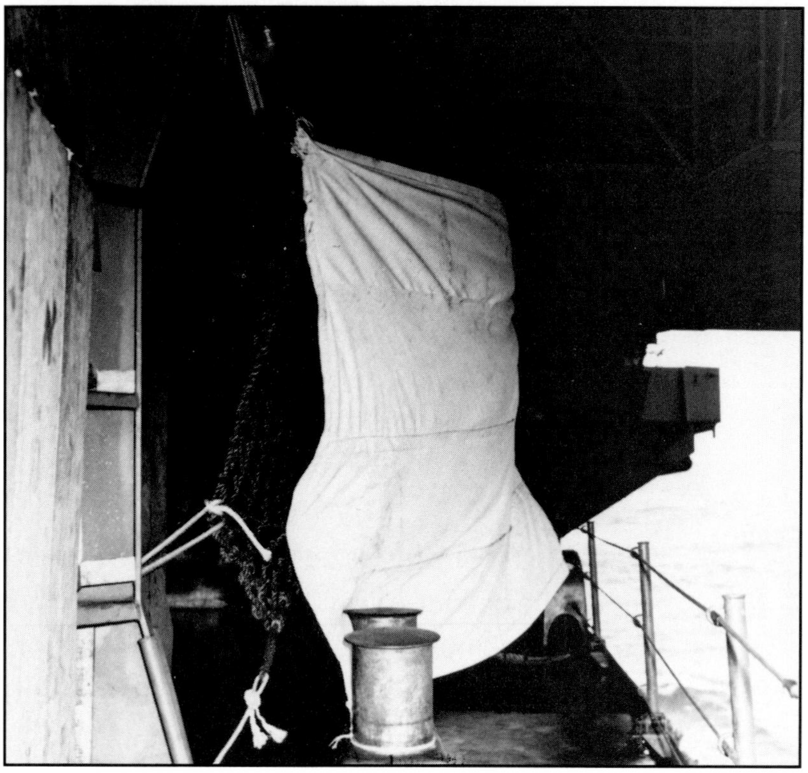

The jury-rigged sail utilized after rudder was severely damaged by torpedo hit. (National Archives Photo)

Two destroyer escorts pull along side, one on the port bow, the other on the starboard bow. Two cables are put aboard us from the two destroyer escorts and they take us in tow. Damage Control is on the job aft. They will try to jury rig a rudder to steer the ship while a crew of men standby to handle this rig once it is in place. They will get instructions from the bridge when and how much to move the jury rig

Chief Bos'n Johnson and Commander Reynolds look on as the Intrepid is under tow to dry-dock at Pearl Harbor. (National Archives Photo)

(rudder). Men who work with canvas are busy gathering all the canvas they can find and begin putting together a huge sail [3,000 square feet] which will go forward at the forecastle to trap the wind and to help move the ship. Some of our planes are moved forward on the flight deck to trap the wind. We are low in the water aft because of the hole in our hull. We flood some compartments aft to drive the stern down further and at the same time raise our bow for more wind entrapment. Crazy! A ship like this with a sail forward and a rudder which is hand operated, but it worked. [After this time all new carriers were built with their bows enclosed below the flight deck — the so-called 'hurricane bow'].

We move at a slow pace with the two destroyer escorts towing us until daylight. By that time Damage Control is reporting on our condition and making final adjustments on our jury-rigged rudder. The sail forward is also about in place. A small crew is aft with the rudder and a small crew is forward with the sail. They both get orders from the bridge on what to do and when. The *Intrepid* is out of action for some time to come, perhaps a little over three months before she will be back in action. All the ships radio us and wish us well.

Captain Sprague alternated our engines and was able, for a short time, to keep a fairly straight course. *Intrepid*, at about 0130 hours, is given a cover to stand by her and to protect her for the time being. My notes say that we had the light carrier *Cabot* (CV-28) and two cruisers with us as well as four destroyers. The plan is that *Intrepid* will try to make Eniwetok lagoon. This group of ships has been designated Task Group 58.2.4. Port screws are turning up full R.P.M.s while starboard screws are getting 1/3 power, so I was told, in order for the ship to keep a fairly straight course at 20 knots.

Out of Action and Home

Between the sail, the rigged rudder, the offset power to the shafts, and the tow ships, the *Intrepid* finally made it to Pearl Harbor for temporary repairs of her battle damage, arriving there on 24 February. About 60 miles from Pearl *Intrepid* launched Air Group Six's 75 planes toward Ford Island. She was put into dry dock almost immediately. By 29 February the huge hole in *Intrepid's* aft hull had been closed in, and her rudder removed. The intent was to steer her by her engines all the way to the navy yard at San Francisco where permanent repairs could be completed. She put to sea about noon with two destroyers as escort, and it became immediately obvious that the huge carrier could not be controlled without her rudder. A call for help by the Captain brought out about eight ships to help her back to Pearl and dry dock again. A strong wind started blowing about this time and it became too dangerous to attempt towing the *Intrepid* back into the harbor, so she stood offshore for the next two days until the weather improved. I made the following notes in my diary:

> Today is 5 March; it is my birthday. I am 20 years old. The *Intrepid* is put back into dry dock. Now the shipyard engineers are going to hang a rudder on the ship the same size as the original one, only this one will be solid, it won't move. They will attach another section to this section which will be moveable. This moveable piece or rudder will be less than 25 % of the original rudder area. Furthermore, its movement would be only an arc of 20 degrees total. This rudder would be controlled from the fantail by a wire setup which would receive pulling power from one of the capstans. Men would have to stand by and from orders received, either turn the capstan right or left which in turn moves the rudder left or right. Truly a jury-rigged rudder. Whether the 100 square feet of new rudder will be sufficient we don't know, but the engineers say it will work.
>
> Very early on the 15th of March the dock is flooded and we prepare to leave Pearl Harbor and, hopefully, the jury rudder will work well enough for us to head to Hunters Point. As is the custom when a ship such as this leaves port, an escort is provided. At this time [16 March] we have as an escort the destroyers *Remey* and *Patterson* tagging along with us. We had some trouble at first, but after a short while everyone got the hang of it and the rudder worked well and appeared to move sufficiently and smoothly. We were able to make 15 knots without trouble and by, at times, using the engines and screws to compensate for the wind and drift. We were finally headed stateside.
>
> The rudderless ship arrives at the outer harbor area of San Francisco and again tugboats pick us up. It is March 22, seven days underway from Pearl. Normally, we make this run in four days. As we get closer to the Navy yard, two more tugs come along side us and the four tugs get us into the dry dock at Hunters Point. As we neared the inner harbor area our planes took off, one after the other, Our AG-6 was headed for another carrier and more Pacific action. I believe they first went to Alameda and joined a carrier there that was headed back to Pearl.... Air Group 6 did a good job for us at Truk and they are now part of Intrepid's history.......

During these days of inaction, and more recently, I had time to do some reflecting on what I had seen and done in these strikes in the Marshalls and at Truk, and these thoughts are a kind of chronicle of an enlisted man's life on board an aircraft carrier in wartime. Most of the following is taken directly from my notes made during these hectic times:

Men are sent over the side to remove scuff marks made by other ships which came alongside after the ship was torpedoed. (National Archives Photo)

> When enemy planes have penetrated the screen and radar did not pick up the approaching enemy planes due to the planes flying low to the water, then torpedo defense will sound. This is a call to all hands to go to their combat positions and guns. The order will come over the P.A. system, the bos'n pipes, the bugler will blow and the gong will sound. The order will be "All hands, man your battle stations!" This will be repeated several times. The gong will sound perhaps ten or twelve times and it would wake the dead.
>
> The call will also go to the pilots and to the airedales [plane handlers on the flight deck]. "Pilots, man your planes!" This will also be repeated. Everyone will run and move and bump into each other and the rule of up on starboard and down on port side will not hold true, as everyone is going in all directions. The planes are usually spotted for take off each night in case they have to take off in a hurry. If they are not spotted, the airedales must move quickly and get the planes in position for take off. Sometimes when this is

Students shoot the sun while Intrepid is in dry-dock #1 at Pearl Harbor having her rudder worked on. (National Archives Photo)

Captain, now Admiral, Sprague passes on the command of the Intrepid and prepares to leave the ship. (National Archives Photo)

being done, the catapults forward will be cleared and planes will be catapulted off the deck, one every one and one half minutes.

Our navigator Commander C.T. Lanman. (National Archives Photo)

The call will go out to the pilots, "Pilots, start your engines"; to the airedales, "Stand clear of propellers." As a plane is ready for take off it will have two airedales, one at each wheel, ready to pull the wheel chocks away. The pilot will rev his engines for a few seconds or so and he will go down the deck and off the ship. As one plane begins to roll, a second plane taxis up to the line and he will do the same and then take off, as will all other planes. Sometimes, before the planes are off the deck, our guns are already in action going after the enemy who is already inside our screen, and many times the enemy is in sight and in range to fire. That is what happens when torpedo defense sounds.

When General Quarters sounds, the enemy could be ten to thirty or more miles away. The same orders will be given to man battle stations and the pilots to man their planes. All hands will remain at their battle stations until the secure order is given. When the flight deck is spotted, the fighters will be in front. Behind them will be the dive bombers and behind them the torpedo bombers. The dive bombers are two men, the torpedo bombers are three men crews. When a strike or a bomb run is planned, the pilots will meet in their ready rooms and are briefed on the targets and the areas, the mileage, etc. They are told what targets to go for and which ones to pass on. The pilots listen attentively to the intelligence officer or to the flight leader as they speak. They are told what to look for and what might be a problem. Will they run into enemy planes or will they surprise them and be over the targets before the enemy can take off against them? This is why many strikes go up early in the morning, so they can catch the enemy by surprise. Sometimes the strike is so quick the enemy does not get off the ground. When a strike is about to take off, the roar of some 75 or 80 plane engines can be deafening.

The airedales admire the pilots and many become real friendly with one another. The airedales worry when the planes take off. They worry while the planes are in flight and they worry until their buddies are back on

Captain William D. Sample reads his orders as Intrepid's new Skipper as acting Captain Commander Gaines looks on. (National Archives Photo)

Captain Sample at his desk would never take command, but was relieved by Captain Bolger before the ship sailed. (National Archives Photo)

deck. There is a bond between the pilots and the airedales. The airedales stand by at take off and they are waiting to secure the planes when they land.

When the flight is off the deck, more planes are ready and brought up from the hanger deck to the flight deck and spotted forward. As the planes land they are very quickly lowered to the hanger deck and rearmed and gassed up. Then the planes forward are moved to the after flight deck and spotted, ready for the next strike off, which could be the same day or early the next morning.

Because of light rains during the night and damp weather, the 20 mm guns, although covered, do not have magazines in them. When I arrive at the gun, which I get to through the Island and through the magazine storage room, I will take two magazines off the rack and bring them to the gun. I insert one load in the gun, the other at my feet to the right. Sometimes I will go back for one or two more and bring them to the gun or as I begin to fire the gun, the man loading the magazines for me will go into the room for more. I never waste any time getting to the gun and many times I was the first to uncover it and begin firing, if I had a target.

The 20 mm battery, which in this case held five 20 mm guns, was identified as #3 tub. These tubs were welded to the side of the outer wall of the Island structure. There was nothing below them but water and the only way into or out of the tub was through the ready room. This first gun in tub #3 was only about 40 or 50 feet from the forward 5/38 mount [a 5 inch gun]. When the 5/38s were firing the concussion was extremely bad on the ears. The 5/38s were on the flight deck and just below them were the 40 mm guns. This made for much noise and concussion when all were firing at once.

The barrels of the 20 mms must be changed about every fourth magazine load or they get too hot. It was the first loader's job to do this. The barrels would be put into a water-filled trough behind the guns on the bulkhead and another cool barrel put into the gun. At one time the barrel had not been changed after I fired the gun for close to one hour and the barrel blew up. It sent shrapnel all over me, cutting my forehead, my nose, my left cheek and my left thumb and hand. At least five 20 mms were on a Japanese plane as he came at us and he was blown apart in a ball of fire, no more than eighty or ninety feet from us. Had the plane continued on he would have taken out this tub and the ten men manning the five 20 mm guns. He was at just the right height and in line with this group of 20 mms. At another time a plane came in the same way but only higher. He smashed into the Island above our heads and the debris fell into our tub and over the side of the ship. At the last moment we all scrambled and ran into the ready room, otherwise none of us would have survived.

I was always thinking and looking around and asking myself what if this or what if that happens. I had a coil of rope hanging in the ready room close by to the hatch (door). Should I ever be unable to get out of the tub it was my intention to tie the rope to the gun mount and go over the side and into the water. I always had a life preserver handy for this same reason. As time went on and engagements with the Japanese planes became an every day happening, the Marine batteries received credit for downing nine Japanese planes and two more probables. Another thought that crossed my mind was to remove or move the gun stops so I could aim the gun lower to help in getting these low flying planes.

I often wondered why all the officers on the bridge stood at the same spot, all together on the bridge, port side of the Island. They would have their arms folded on the railing and observe the flight deck activity. The Japanese must have noticed this also because many a suicide plane headed right for the spot. Look at any photograph of a Captain or officer on the bridge and he will be in that spot with his arms folded on the top of the shield or railing of the bridge.

Helen Evedith tries to get crew member up on the stage in California 25 April, 1944. (National Archives Photo)

Danny Kaye and Dick Powell on stage on the Intrepid on 29 April, 1944 at Hunter's Point Dry-dock. (National Archives Photo)

Dick Powell at the mike 29 April, 1944 in dry-dock at California. (National Archives Photo)

Intrepid afloat in dry-dock, while checking rudder to see if it responds properly.. (National Archives Photo)

While waiting for the enemy who may be anywhere from five to thirty miles away, the men talk and pass the time as though they were at a picnic or sitting in a bar room somewhere. They are relaxed and anxious to start firing. At times, they make bets on who will see the first plane approaching the ship. Some men who just got off a watch may sit on the bench in the ready room and more than once I have seen a man actually fall asleep while waiting for something to happen.

I spent many hours making notes in my diary and for a period of time I made a map of the ship's position for each day of her travels. Having a fair understanding of navigation, I was able to draw this map laying out the course and drawing in the different enemy islands and location. Each day I would get the ship's exact location (longitude and latitude) and record them and at a later time, using these positions, lay out that particular course on the map. The map was 2 1/2 feet by 3 1/2 feet in size. Some of my notes would be written while on duty as orderly at the Captain's or Admiral's cabins; other times I would write late at night while everyone was asleep.

Since the night of the torpedoing of the Intrepid (0011 hours into 17 February 1944) I never slept in my bunk. While in combat areas, I slept under the guns or on the ready room bench, if the weather was overcast as it was many times. The days could be hot, sunny and sticky, but at night it got chilly and damp and wet. I used to use my life preserver as a pillow and a poncho as a blanket.

Other means of passing the time were card games, poker games, rolling dice, reading, studying, sleeping or just plain talking. I spent my free time with my diary, adding or deleting information or working on my combat map. Some of the things that got written about were what happened that day in regards to the strikes that took off. How many planes went up, how many did not come back? What type of bombs did the bombers carry? Were they incendiary or explosive? Were they 500 pounds or 1,000 pounds? Was any returning plane shot up? Was any pilot injured? This is what made my day or night in passing time. I would always try to find out what ships were with us, and at what position they were in, in relation to us. How many miles away were the battlewagons and carriers and destroyers? How many destroyers made up the outer screen all around the big ships? My mind and my pen were always in action as to who got hurt and how bad things of this nature made up what I tried to put down on paper. When a plane goes down in the water, who is the pilot, what carrier is he on, what destroyer picks him up and how many gallons of ice cream does the destroyer get for returning the pilot to his ship? What else does the destroyer get? Perhaps a crate of apples or oranges or some movies.

Then there is the question every service man who has been in action is asked and that is: "Were you scared at any time?" the answer is "Yes." Everyone thinks about his safety at times; that is only natural. Whether the ship is under attack or burning or has taken a torpedo or a suicide plane — yes, we were all scared at one time or another. My most frightening time was when the Japanese plane hit the Island over my head and I thought sure enough this was it and I had bought it. I think at that time I was the first man to get into the ready room.

What are some of the actions of the pilots and the planes flying CAP (Combat Air Patrol)? How do they maneuver in combat? The carrier planes usually fly just above the usual Japanese flight pattern. The American planes try to stay above the attacking planes and then drop down behind their selected target and line up with him and shoot him down. Many times the enemy doesn't know we are there until he is hit and shot down. The kamikaze pilots, after a while, got smart. They would head for their target very high up — as high as 20,000 feet. At this height radar sometimes would not pick them up. They then would go into a steep 90 degree dive and head straight for the superstructure of a ship or the center of the flight deck of a carrier. This happened four different times on the Intrepid .

As the carrier planes return, there is much discussion as to what took place, what did they hit and who got shot down. This discussion always takes place even before the pilots return to the briefing room to officially report on their strike or raid or bomb run. It is then that the flags are painted on the side of the Island showing the sinkings, the ships damaged, the planes shot down. The pilots add the flags to the side of their own planes if they shot down a plane or hit a ship.

The 5/38 mounts had their score board as did the 40 mm mounts. The 20 mm batteries also had their score up on the bulkhead or in the ready room. The brains of the ship, the Captain, Admiral, flight officers, landing officers, signal people, gunnery officers, are all up on the bridge or signal bridge, on the port side of the Island and all leaning on the railing with their arms folded on the railing. The flight officers and the landing officer will communicate with their fingers to one another on how many still to land or how many did not come back. Instructions are given by the Captain and air officers to turn into the wind and land planes or to the engine room calling for more speed to close the gap between the ships. When landing operations are taking place, sometimes the carriers fall behind the others and must catch up when the planes are aboard. The gunnery officer watches closely

Intrepid takes many sailors with their gear on board at Alameda, California 8 June, 1944 for transport to Pearl Harbor. (National Archives Photo)

Along with sailors, the Intrepid transports other cargo to Pearl Harbor. (National Archives Photo)

A closer look at the various items transported by the Intrepid to Pearl Harbor. (National Archives Photo)

A wedding on Intrepid's deck on 9 May, 1944. Patterson and his new wife, Ann with Chaplain Lieutenant Commander Donald Kelly, a Catholic priest from Independence CVL-22. (National Archives Photo)

the tracer fire and when a plane is shot down he knows pretty well what gun or battery brought that plane down, and he will say over the P.A. system, "Nice shooting gun so and so, or Battery 1—2—3, whoever did the downing of the enemy plane. The gunner being addressed will turn and wave to the bridge. A little confidence and praise does wonders for a man's morale. Things quiet down and the Captain will give the order "Secure from General Quarters" or "Secure from torpedo defense." The ship then goes about its normal duties. The men relax. Maybe we might even get to eat something.

Many in *Intrepid's* crew were given passes and leaves while the ship was in drydock for repairs. *Intrepid's* skipper, Captain Thomas L. Sprague, was promoted to Rear Admiral and, on 29 March, turned over temporary command of the carrier to Commander Richard Gaines, her Executive Officer. In the interim period, until repairs are completed, both Gaines and then Captain William D. Sample were alternately in command. Also during this time, Senator Harry S. Truman came aboard as part of his inspection tour of the Navy Yard. One of the highlights of the period was on 2 May when *Intrepid* held her first wedding (James C. Patterson, RM2/C to Ann Gargas, S1/C [WAVES]).

A Senate inspection team on a P.C. boat next to the Intrepid at Hunter's Point. Harry Truman is second from right. (National Archives Photo)

The non-paying passengers disembark at Pearl Harbor 13 June, 1944. (National Archives Photo)

The big carrier was in drydock #4 at Hunters Point until 25 May, 1944 when she was floated clear and moored. All her underwater repairs had been completed but there was still work to do inside the hull. On 30 May Captain Joseph Francis Bolger assumed command of *Intrepid.* On 3 June most of the internal work was completed and *Intrepid* put to sea for post-repair trials. The next day she returned to Hunters Point for completion of minor alterations and repairs. On 6 June she was sailed across San Francisco Bay to the Naval Air Station at Alameda, and three days later at 1415 hours set a course for Pearl Harbor carrying passengers, aircraft, vehicles (trucks and tanks) and miscellaneous cargo, Captain Bolger commanding. From the time that *Intrepid* took the torpedo on 17 February, she had now been out of service for nearly four months.

The four day passage to Hawaii was not uneventful. Every day the carrier test fired her 5/38s and her 40 mm and 20 mm guns. On 12 June, a day out of Pearl, the helm "froze" and the bridge lost steering control and the reduction gears of the No. 2 propeller shaft failed. Steering control was eventually regained but it was obvious that *Intrepid* would have to go into drydock again. She was assisted by tugs into the harbor on 13 June where repairs began on her steering; on 21 June *Intrepid* was drydocked for repairs of the reduction gears and the next day was floated and returned to Ford Island.

On 22 June she received on board the planes and personnel of Air Group 19, a temporary assignment, and the following day was underway with a two destroyer escort for Eniwetok Atoll. There was no time to be wasted; the striking power of the big carrier was badly needed. But this was not to be for there were still problems with her steering controls. On 30 June the atoll was reached and by 1900 hours she was anchored in the lagoon. There she would stay until the late afternoon of 4 July when she was ordered underway to Pearl Harbor with 551 Army, Navy and Marine battle casualties on board, escorted by three destroyers. While at anchor in the lagoon, *Intrepid* catapulted off the entire Air Group 19.

By 11 July *Intrepid* was once again moored at Ford Island in Pearl Harbor where she disembarked her passengers and repairs were again started on her balky steering mechanisms. She entered the drydock on 22 July for below waterline repairs which were completed by 25 July when the ship was floated but remained moored to the drydock. For the next four days she took on fuel, bombs, ammunition and stores and on 29 July put to sea for post-repair trials. More problems: the repaired reduction gear train overheated at full power turns and *Intrepid* was forced to return early to Pearl. For the next two weeks, until 16 August, repairs, sea trials, and gunnery and flight practice operations were conducted, until the mechanical and operational problems were finally overcome. On 10 August, at sea, Air Group 18 led by Commander William E. Ellis, was landed aboard and reported for duty.

The War Goes On

During the months that the *Intrepid* was undergoing repairs, the Navy had been active, immediately after the Truk strikes, in bombardment and aerial strikes in February against Saipan, Tinian and Guam in the Marianas, some 600 nautical miles northwest of Truk. These islands were assaulted by U.S. Army and Marine troops in June and July, the last, Guam, being secured on 10 August 1944. The Mariana Islands are 1100 nautical miles from the Philippine Islands, General MacArthur's ultimate goal. Carrier task groups then hit the Palau Islands and Yap in March, southwest of the Marianas.

Through the winter, spring, and early summer months of 1944, Allied troops under MacArthur's command had been pushing back the Japanese forces along the north coast of New Guinea in a relentless coastal-hopping drive westward. By the end of July MacArthur's armies controlled the far western end of New Guinea, less than 600 miles from the Philippine island of Mindanao. For the Carrier Task Force's part in some of these operations, consider the attacks on the Palau Island group (note that our *Intrepid* did not take part in this action; she was undergoing repairs at San Francisco):

> ***Wednesday, 29 March and Thursday, 30 March.*** Several days earlier the Palau group of islands are photographed by one of MacArthur's Liberators [B-24]. The spy plane was seen by the enemy and it is believed that what ever naval ships were there then, left, as the first strike by the carrier planes reported no combat ships to speak of, were seen. Some of the escaping ships were spotted by U.S. subs on the 29th; they were headed northwest to more friendly waters. On the 30th, carrier planes go northward to search for the enemy ships, but none are found. On the 28th, late in the day as the carriers prepare to launch a strike early next morning, at least two enemy torpedo planes attack a section of the Task Force. CAP promptly shoots them down. No damage to the ships.
>
> The strike next morning is met by some thirty enemy planes already in the air. Within minutes at least ten enemy planes are shot down. Many ships are in the lagoon and they are getting up steam to escape. The outer reef of the Palau group had only three breaks in it to allow ship passage and these were so tight that ships had to exit single file. The islands of Palau had much jungle growth and it began right at the water's edge. This hid gun emplacements and even smaller ships. The torpedo planes went in, not with torpedoes, but with mines and these were dropped all over the area which stopped ship movements cold. Fighters escorted them, strafing the area around the drop points. It was reported that some fifty ships, perhaps more

which had been hidden close to shore, were trapped in the lagoon. Then a strike of bombers and torpedo planes went to work on the trapped shipping.

Many strikes were made, each carrier making six runs that day. The Japanese then began to move ships to shoal water and even beached some of them. One strike sank three ships and damaged seven others. Another strike sank three more and damaged six others. Another strike hit six more ships. As each strike came over the lagoon, ships already hit were hit again. Three ships got out of the lagoon and were promptly sunk by the big guns of our surface force standing out off Palau. Two planes were hit over the target but the pilots managed to get them back to their carrier. These two planes received damage on their attack on an enemy cruiser and two destroyers. The cruiser was left in flames, one of the destroyers hit tried to re-enter the lagoon and was sunk close to that entrance. The other destroyer got away in the open sea. A little later on, eight U.S. torpedo planes are on a search mission and they spot this destroyer. The eight planes, each from a different angle, attack the destroyer. Eight torpedoes go toward the destroyer. In a few minutes there was a terrific explosion and when the area clears nothing is seen of the destroyer. Most of this action was on 29 March.

Day two strikes found many new planes on the enemy airfield. Where they came from is anybody's guess, as they were brand new. On this day one Air Group alone got 29 planes. Another group had 25 shoot downs, with eight more damaged and two more probables. Many times our planes are hit, sometimes bad, but most times the pilots somehow get them back to the carriers. I've watched many a plane astern headed for the flight deck. They would weave and dip and be anything but in a proper landing pattern. On they would come, some straight to the barrier, some would be jerked to a stop and the plane would break in two. Others for some reason would turn sharply and pile into the gun tubs or hit the Island. It was amazing how many of these crack-ups had their pilots walk away unhurt. Then there was the time one plane still had his belly tank under him — he had not jettisoned it before coming on board. As he hit the deck, the spare gas tank fell and in a few seconds the plane was a raging inferno. After that the pilots began to drop their spare gas tanks over the target area and they acted like Napalm bombs setting fire to where ever they landed.

Another downed fighter in the water was observed trying to get his life raft out as the plane began to sink. Another plane over him somehow got his raft loose and dropped it to the pilot below as the plane went under with the pilot standing on one wing. He was later rescued. Then there was the pilot who went down in flames and no one saw him get out of the fiery plane. The pilots reported when they got back that perhaps the pilot was lost. He was put on the missing list. Another plane landing on deck hit a gun battery with one wing. The wing broke loose and went far forward up the flight deck; the plane broke in two right behind the pilot's seat. Medics rushed to the still intact front of the plane to rescue and help the pilot. Before they got to the plane the pilot got out, walked up the deck towards the medics, waved and said, "Hi, fellows." He was not injured in the least, except for a slight scratch on his face.

The following day a message came through about the plane which went down in flames and no one saw the pilot get out. One of our smaller ships, perhaps a mine-layer off shore, had seen the plane go down, went to investigate, and rescued the pilot who was not hurt. This same ship also rescued one other pilot. It would be impossible to log all the acts of heroism which took place because there were hundreds of incidents which took place on the carriers, in the water and at the target areas.

After the raids on the Palaus, the force bombs Yap and Ulithi and Woleai. Ulithi was taken without a struggle, later, on 22 September. Ngulu, 40 miles south of Ulithi, was also taken. At Woleai not even AA fire greeted the planes when they struck the island. It was bypassed.

MacArthur's Operation "Reckless", the seizure of Hollandia on New Guinea's north coast, began on 21 April and was a push over. The Task Force, twelve carriers strong, then moved back to Truk to hit it on 29 and 30 April. The Task Force on several occasions, had enemy snoopers after leaving Hollandia, so you can bet the Japanese at Truk know we are on the way back and they would be waiting. Reports state that the Japanese have close to 200 planes on the airfield. One enemy Betty which was following the task Force was shot down as were some six other Japanese planes sighted on the way back to Truk. Eight Japanese were picked up out of the water - one officer and seven men. These men would be grilled later by staff officers.

The first strike [on Truk] hit the harbors and airfields. The second strike hit docks, buildings, gun emplacements, and again gave the airfields another pass. Paran and Moen islands each had airfields and they were hit. Parts of the inhabited island of Dublon were also hit. A seaplane base west of Dublon was demolished. A fish cannery was also flattened. Eten Island, another small airfield surrounded by anti- aircraft guns, was bombed to pieces. The first day, 29 April, saw sixty enemy planes shot down plus sixty more wrecked on the airfields.

The second day was a repeat of the first. What was hit on the first day was again hit on the second day. Many planes went down in these two days but the submarine *Tang* rescued twenty-two air-crewmen from the waters off Truk and other islands with the help of an OS2U Kingfisher plane off one of the cruisers. Truk was now a shambles. Even if the Japanese try to rebuild Truk again, it would take them many months to do it and their work would be a waste of time because we are coming back again, and then again, to blast the rock of the Pacific, which now is just a pebble, and we intend to keep it down-sized.

The months of April and May 1944 saw the fleet carrier force roam westward along the coast of New Guinea in support of MacArthur's landing operations, and toward the Marianas and Palaus, hitting enemy airfields to protect these land operations. During these months *Intrepid* is undergoing major repairs at Hunters Point in San Francisco Bay. In my diary I recorded the following notes:

> ***June 6.*** It is D-Day in Normandy, France, as allied expeditionary forces under General Dwight D. Eisenhower land on the beaches at Normandy, France.
>
> ***June 7.*** On this day, the 14 inch guns once silenced at Pearl Harbor, are in action again. Risen from the dead, ships at battleship row in Pearl Harbor after the sneak attack of the Japanese on December 7, 1941, is the U.S.S. *Nevada* as she bombards a Nazi position at the Normandy invasion. The badly crippled battle wagon was brought back to modern fighting trim and this is her first action. "Welcome back U.S.S. *Nevada*. Give 'em hell!"
>
> ***June 10.*** At this time part of the fleet leaves the Marshall Islands for an assault on Saipan. The force headed for Saipan will be rather large. It will consist of four battlewagons, two fleet carriers, fourteen destroyers, three light cruisers, one heavy cruiser and three mine sweepers. They will be joined by a second unit. A third unit will also be in the area less than one day's travel away. This third section of the fleet will be the largest of all. It will have seven battleships, five carriers, 65 destroyers, heavy and light cruisers and a unit of the submarine force also goes along. Fully constituted this invasion fleet will be made up of fifteen carriers carrying some 800 planes. Bringing up the rear are supply ships and transports loaded with some 127,000 troops. It was reported that the men manning these ships numbered in excess of 99,000. Part of this fleet heads for Saipan and Guam, 1300 miles away.....

This enormous force, designated Task Force 58, is at first commanded by Admiral Raymond Spruance but soon command is turned over to Admiral Bill Halsey. Admiral Pete Mitscher, the Carrier Commander, remains in that capacity. The task now at hand is to hit the Marianas and destroy all shipping and enemy air power at and on the islands. The islands to be targeted are Saipan, Tinian, Rota and the big island to the south — once owned by the United States — Guam.

The first strike is planned for 12 June but the force was detected by enemy snooper planes on 10 June and Admiral Mitscher orders the force to make best speed to close the distance between the force and the launching area. On the afternoon of 11 June, some 200 miles from the target, Admiral Mitscher orders the strike to begin. 200 planes of the first strike leave the carriers to hit the three big islands of Saipan, Tinian, and Guam. The strike made that afternoon destroyed 124 enemy planes.

June 12. On this day as an amphibious assault on Saipan in the Mariana Islands is approaching and the action to be known as the Battle of the Philippine Sea begins. Part of this action will also later be called "The Mariana Turkey Shoot." On the first day two U.S. cruisers, the *Cleveland* and the *Birmingham* will be hit.

June 14. On this day just prior to the invasion, Saipan is again bombed. The bombing left the sugar cane fields ablaze and flattened. The sugar cane refinery is bombed and is leveled. Saipan, Tinian and Guam will be bombed repeatedly and all three invaded. The airfield at Aslito on Saipan is completely wrecked and torn apart. Many Japanese planes were wrecked on the ground.

June 15. ..... After three days of bombardment, the troops land on Saipan on this day. Second and Fourth Marine Divisions and the 27th Army Division. We invade Saipan and the battle begins for the Marianas and turns into the Battle of the Philippine Sea. The code name for this Saipan operation is "FORAGER." On this day another terrifying setback for the Japanese as our B-29s hit Yawata, Japan. Their target was the Imperial Iron Works. It produces 25% of the Japanese steel. The Japanese anti-aircraft fire was heavy and four of our very expensive B-29s are shot down over or near their target area. The B-29 is half again larger than the Flying Fortress. A new strike force was created for these B-29s and it is named the 20th Air Force. The planes took off from one of our China bases. On the return flight home, the planes hit the Shawa Steel Works at Anshan, Manchuria.

June 16. Two units of our force headed by Admiral Clark at the Saipan invasion break away and head for Iwo Jima. They bombard Iwo Jima real good. The raid was made to hit the Japanese airfield and the planes there to keep them from coming at the invasion force at Saipan. The ships do their bombardment and quickly turn and head back to the fleet at Saipan which is now being invaded. We have two units of the Japanese fleet headed for Saipan. Saipan receives a massive bombardment on this day. The bombardment began on the 12th of June and is increased this day, June 16th. The troops have been ashore on Saipan for 24 hours.

The two carrier groups that had been hitting the Jimas turned south leaving a wake of destruction behind them. All day on 17 June the force headed toward Guam at high speed to join the rest of Task Force 58. The Japanese fleet was reported out of port and heading for the Marianas, drawing in units of its navy and growing in size at it approached the islands. This could be what everyone was waiting for; the final showdown between two mighty navies and air forces. The Jap fleet, commanded by Japanese Admiral Ozawa, was spotted and reported by a submarine on the morning of the 18th — nine carriers escorted by battleships, heavy and light cruisers, destroyers, oilers, the whole works. They were heading into the Philippine Sea west of the Marianas. That afternoon Admiral Spruance's huge Fleet was spotted and reported by enemy snoopers. Instead of heading west toward the approaching Japanese fleet, Spruance ordered an eastward course to cover the now three day old invasion of Saipan. The enemy still had plenty of planes on Tinian and Guam, just south of Saipan so the Task Force was now sandwiched between those bases and the oncoming Japanese fleet to the west. On the morning of 19 June fleet radar spotted "bogeys" - enemy planes - over Guam. It turned out to be a couple of dozen Zeros heading north to Saipan to help their embattled countrymen. More were taking off and Admiral Clark ordered Hellcats to Guam to do battle.

So began the first day of the Battle of the Philippine Sea, which soon became known among Navy airmen as the "Mariana Turkey Shoot." Within minutes after launching at about 0745, planes from the *Yorktown, Belleau Wood* and *Hornet* were strafing airfields on Guam and splashing several bogeys on the way back to the carriers. Two hours later radar picked up the leading elements of the enemy carrier fleet's aircraft at 24,000 foot altitude coming in from the west. The enemy fleet itself (identified as the First Mobile Fleet by the Japanese) was over 300 miles away and these planes were on a one-way trip, presumably heading for a landing on Guam after their attack on the U.S. Task Force. Admiral Mitscher smelled the big event coming and ordered every fighter under his command into the air. All SB2C's and TBM's were loaded with bombs and virtually every other kind of ordnance available and sent to Guam to do as much damage as possible. Radar was now updating the incoming enemy strike: 100 miles out and approaching at 200 miles per hour; 75 miles out; 60 miles out; and then the two opposing forces made visual contact. The enemy strike was in two groups, the first about 40 planes, but the second was reported as "several hundred." The Japanese planes were below the Hellcats at about 18,000 feet and the U.S. fighter pilots rolled over their F6s and power dived from 25,000 feet straight into them. The enemy planes were fighters; the enemy bombers were still lower. Now the carnage began: The Hellcat pilots blasted away at enemy fighters between 20,000 feet and the ocean's waves. Zeros (Zekes) and Tonys were being flamed left and right, high and low; fighters at near full throttle were approaching each other in head-on runs at approach speeds of over 650 miles per hour. Hellcat pilots were flaming two, three, four, five enemy planes each - many Aces were being made this day.

More waves of enemy planes were being plotted by Task Force radar - raid 5, raid 6, raid 7, and still coming in. Some Japanese torpedo bombers broke through the fighter screen but none got close enough to drop their "fish." A brief respite at 1100 hours as the air was finally cleared of enemy planes. The calm was used to launch Hellcats, Helldivers and Avengers to Guam. They dropped 21 tons of bombs on Guam airfields this day. Then enemy raid 8 was spotted about 100 miles out and heading for the Task Force. All available Hellcat pilots were "scrambled" to contact this force; visual contact was made at about 1140 hours — 60 enemy bombers and a few fighters. More carnage was wrought by the Hellcat pilots on this latest raid. But some of the bombers broke through the fighter screen. Eight of them started a run on a battleship and all eight were promptly shot down by the battlewagon's immense fire power. Small groups of about 30 enemy planes continued to attack the Task Force into the afternoon; not one got through. The enemy's carrier planes were being decimated, but a search flight of fighters and dive bombers from the *Yorktown* sent westward over 300 miles failed to locate the enemy fleet. Two U.S. submarines were luckier, though; they did find the enemy fleet this day and sent two Japanese carriers to the bottom. The *Shokaku* was torpedoed by the submarine *U.S.S. Cavalla,* and the *Taiho* was sent to the bottom by the *U.S.S. Albacore.* The Navy's intelligence concluded that the enemy had seven carriers left in its entire fleet.

The aircrews of the Helldiver and Avenger raid sent eastward to Guam now less than 50 miles away, were having a field day. Escorting Hellcats swept the air of enemy planes as the 2Cs and TBMs destroyed airplanes and everything else in sight on the ground. By late afternoon the air battle was over and that evening Task Force 58 headed west in search of the enemy fleet. The final score for this day of 19 June varies depending on whom you read, but was in any case extraordinary: Of an estimated 430 carrier-based enemy planes sent against the Task Force, about 370 or 85% were destroyed although some records put the total number at over 400. An additional 50 or so island-based enemy aircraft were also destroyed. Sixteen Hellcats were lost in the day's combat but some of the pilots were rescued; a total of six Avengers and Helldivers were lost to flak hits over Guam. Forty-nine T.F. 58 air crewmen lost their lives this day. Tokyo Rose, the following day, reported that two American carriers had been sunk and 300 American planes destroyed; somehow she got the combatants reversed.

The next day, 20 June 1944, dawned rainy with a low ceiling. The *Yorktown* got off a search group of eight Hellcats at about 0600; before 0800 they had intercepted and shot down two scouting enemy float planes, so the Japanese fleet was close at hand. Three more enemy planes, this time from a carrier, were "splashed" soon after. Then the pay-off: in early afternoon a scout plane (from the *Hornet)* spotted the enemy fleet and reported its position (270 miles west), its speed, direction and make up. The force was in three large groups: (1) the carrier *Zuikaku* with four battleships five cruisers and eight destroyers; (2) a large carrier with its battleship, cruiser and destroyer escorts; (3) two small carriers, ten destroyers and two oilers. The report missed three more carriers sailing with this force. The decision was made to go after the enemy fleet, even though it would surely mean a dangerous night recovery of the planes. Landing on a pitching, rolling carrier deck was challenging enough in broad daylight; at night, it was courting disaster. There was no choice for this was the greatest opportunity of the Pacific war to deal the enemy a significant - perhaps decisive - blow.

Beginning at about 1615 hours Task Force 58 launched 216 planes to strike the enemy, now about 300 miles to the west; a long flight out and a long one back. Two and one-half, boring, droning hours later at about 1840 hours the Task Force's 200 plus planes "tallyhoed" (sighted) the enemy fleet. There were four groups rather than three, but only three groups included carriers. They had flown through the squalls and weather conditions were now excellent; but the sun would set just after 1900 hours - a half hour of surface daylight remaining. Seventy or eighty Zeros were flying CAP over their fleet as Helldivers and Avengers screamed in for the kill, foregoing the usual coordinated, text book attack. Eight torpedo-carrying Avengers and 15 Hellcats, each with two 500 pound bomb, pounced on one group containing two medium displacement carriers (the *Jinyo* and the *Hiyo*) and a large carrier (the *Ryuho*). Thirteen Helldivers started a dive run on the heavy carrier *Zuikaku*, the last carrier survivor of the fleet that had attacked Pearl Harbor two and a half years ago. Planes from all the Fleet carriers chimed in on these groups and on a third group with carriers. The Helldivers got at least seven direct hits on the *Zuikaku* and several near misses; fires were started at the "island", on the flight deck, and in the hanger deck below. The sky seemed like a solid wall of flak being thrown up by the enemy fleet.

The carrier *Ryuho* probably got at least two torpedo hits but did not go down. A TBM was flamed by the carrier's gunners and went down with all three men. Four torpedo planes, carrying bombs rather than torpedoes, scored direct hits on a light carrier, (the *Jinyo* or the *Hiyo*) as bomb-carrying Hellcats also made direct hits on the same carrier. A TBM was hit and destroyed by flak fire from this ship. The attack didn't last long; by 1900 hours the Hellcats, Helldivers and Avengers were heading home. The final score for the attack would be the sinking of the *Hiyo*, and serious damage to the *Zuikaku* and three other carriers. In addition, two oilers were sunk, and about 65 enemy airplanes shot down. But American losses were just beginning.

The ships of Halsey's Task Force were over two hours flight time away, and all in pitch blackness. Navigating to this pinpoint in the Pacific would be demanding enough; landing safely in the dark would stretch every pilot to the limit of his abilities. And there was the fuel problem — most of the planes had barely enough gasoline left to make it back to the carriers; some of the 2C Helldivers didn't even have that much left and their crews knew they would have to ditch short of the fleet. Many did. So all pilots did what had to be done to conserve fuel: They throttled back until they were flying just above stall speed — dangerous in daylight, compounded in darkness where there was no horizon and pilots were flying on instruments alone. And the fuel/air mixture was leaned out until the big 2,000 horsepower engines were starving for gasoline and began to overheat to dangerous levels, spewing white-hot sparks from their exhausts. Cylinder heads were known to melt under these brutal conditions.

A little over 60 miles from the fleet the leading pilots heard the homing beacon and adjusted their courses slightly to bring them over the carriers. Some night fighters were sent up by the fleet to rope in some errant planes heading in the wrong direction. Almost home.

The Task Groups were separated about 15 miles from one another. At about 2000 hours in total darkness the carriers turned east into the wind and bolted to 25 knots speed to recover the incoming flight. Each of the twelve carriers in the Task Force turned on a few dim and hooded lights to outline the boundaries of their flight decks — barely visible until the landing pilots were almost on top of them. The incoming pilots were told to land on any carrier since none had enough fuel left to allow them time to mill around in the blackness to identify their home base. And the chaos began. With only a few gallons of gasoline left in their tanks planes were edging out others in the landing patterns, coming in high, low, fast and slow, missing arrestor hook wires, piling into the crash barriers. Decks of most of the carriers were fouled with the debris of crashed Hellcats, Helldivers, and Avengers; planes were running out of fuel and crashing into the sea all through the extent of the huge fleet.

A first rate disaster was not just in the making, it was taking place. Admiral Mitscher then ordered the fleet's search lights to be turned on and directed skyward as a beacon for the returning planes. The *Yorktown's* Captain Jennings then ordered his carrier's flight deck to be bathed in the light of a powerful cargo lamp, pre-rigged by *York's* flight deck operations veteran Pappy Harshman for just such an occasion. Air Plot sent out a message to the returning flight that the *Yorktown* was lit up. Admiral Mitscher ordered more lights turned on in his carrier groups and the escorting ships lobbed star shells into the air to light up the force. Soon virtually every ship in the gigantic Task Force turned on every available light and the ships of the fleet became a sitting duck for any enemy submarine that may have spotted them. There were no sub contacts that night however. With the carrier flight decks bathed in light, some of the pressure was off in at least the actual approach and touch down of the planes on the deck. But everyone was dangerously low on fuel, some planes burning their last drop just as the hook trapped a wire. Frequently two planes would be only yards apart on their final approach to the flight deck; the LSO (Landing Signal Officer), who signaled the pilot with paddles telling him where he was in relation to the flight deck and exactly when to cut his engine, was the man of the hour. LT. (j.g.) Dick "Dog" Tripp, *Yorktown's* iron-nerved LSO would bring in the lead plane and signal engine cut to hook on to the last or next to last wire on the deck; he would then concentrate on the second plane and order engine cut to catch the first or second wire. The two planes landed almost simultaneously. LSOs averted one disaster after another this night and earned their meager pay many times over.

But returning planes were still running out of gas and splashing all around the brightly lighted ships; the aircrews couldn't even see the water as it came up to engulf them in the darkness. The planes simply crashed to a sudden stop from 70 or 80 miles an hour and everyone scrambled to get out — if they were still conscious — before the planes sank. Cruisers and battleships launched SOCs to search for downed pilots and air crewmen. Destroyers and submarine *Tang* picked up many of the men from these crashes within minutes. Other planes came sliding in on their bellies with their wheels locked up from battle damage. Many of these crashes caused fires on the deck and so destroyed the planes that they were simply pushed into the ocean to clear the deck for other incoming planes. The *Yorktown* alone took aboard a total of 23 planes — 16 from other carriers plus seven of her own. About 2200 hours a Japanese Zero appeared in the landing pattern of one carrier(!) but eventually was lost by radar. By 2230 hours it was all over except for finding and picking up downed pilots and aircrews. Admiral Mitscher turned the Task Force around and went back over the route taken by the returning planes. The search went on for two days which allowed the Imperial Fleet to get further away. 84 planes ditched in the water. 16 planes were downed by enemy anti-aircraft fire over the ships as the attack took place. Of the 84 water landings, 48 men were lost. But Japanese Naval air power would never recover from these two days of devastation.

Saipan's airfield will be secure and in use by our planes on the 18th of June. The island will not be totally secured until July 9. In the first week of July (the 6th), the Japanese troops will launch a banzai charge in a last effort to push back the Americans. It will fail. The Marines with their mortars and flame throwers and automatic weapons, will slaughter this last effort of the Japanese guard. By June 22 Army P-47s land on Saipan's repaired airfield. On July 6, 7 and 8, the Marines and Army troops concentrate on the last Japanese holdouts in their bunkers and dug-in emplacements.

The banzai charge by the Japanese on July 6 is commanded by Admiral Nagumo (beaten at Midway in 1942) and Lt. Gen. Saito; both will be killed in this charge and last defensive action. It took months to get at the few Japanese who were

A gasoline barge delivers the U.S.S. Intrepid aviation gas 2 July 1944. Note escort carrier in photo, the ship is at Eniwetok Lagoon, just west of the main Marshall Islands. (National Archives Photo)

The Intrepid takes on board wounded men at Eniwetok Lagoon 3 July 1944. The wounded are transported by LCTs to the ship. (National Archives Photo)

The Intrepid enters Eniwetok Lagoon 4 July 1944. Many other U.S. ships are anchored here with the Intrepid. Photo was taken from the Intrepid's aft starboard quarter. (National Archives Photo)

Army personnel (L to R) Sgt. Michael Walzanski, Sgt. Renoldo D'Antonio, Sgt. Charles Saladine and Sgt. 1/C Jos. Casha discuss a captured Japanese Flag 6 July, 1944. They are passengers on their way to Pearl Harbor. The Intrepid has left Saipan, they have 500 wounded men on board and are headed for Pearl Harbor for repairs. (National Archives Photo)

holding out in their concrete- and steel- reinforced bunkers of which they had many in the hills. The capitol of Garapan is in ruins - all buildings are flattened. Any and all targets above ground were hit and pounded flat by the big guns and the plane strikes. Japanese losses are high. the U.S. forces of 65,000 will have 3,400 killed and 1,300 injured. The Japanese force of 24,000 will have 2,000 taken prisoner and the rest will die.

Saipan is about 14 miles long and has some high ground and hills at the north end of the island, which is where the troops had trouble with the last Japanese holdouts in their bunkers. Saipan is only about five miles wide. Population is said to have been 50,000 at one time. There was a railroad line which circled the island close to the shoreline.

The combat ships in Saipan harbor at the time our troops landed, were one battleship, two cruisers, and five destroyers. During the night of 13 June the Japanese submarine RO-36 surfaced directly in front of one of our destroyers. He is sunk before he can dive. The same night a Japanese cargo ship will try to sneak out and it will also be sunk by a patrolling destroyer. On 20 June the enemy carrier *Hitaka* was sunk by aircraft in the Philippine Sea area west of Saipan. The island

 A SB2C-3 of the VB-13 Squadron flies over the Intrepid as it is steaming towards Pearl Harbor 6 July, 1944. (National Archives Photo)

of Saipan with all its greenery and palms, was so pretty and now it is a flat, barren-looking place after all the shelling and bombing. A sugar cane factory, some oil tanks and even the sugar cane fields are all leveled. The railroad and the trains were demolished. Trucks and wagons were hit. Anything that presented a target was destroyed. Enemy troop concentrations, their barracks and airfields were all hit by planes who flew rampant over the island. For some time the sky will be dark with smoke.

For days there are many Japanese bodies floating in the water. They include women and children. Some are not clothed. On Saipan, two large piles of Japanese bodies are bulldozed into mass graves. Thousands of bodies are buried this way.

Between Saipan and Guam is Tinian, separated from Guam by a narrow (less than four miles) channel. At first we were going to bypass Tinian and concentrate on Guam but the plan was changed. The main activity on Tinian is sugar cane. This island, like all the others, is pounded by the big guns. On the first night (24 July) the Japanese rush our men (2nd and 4th Marine Divisions) on White Beach to try to drive them back into the sea. The Japanese were cut to ribbons by the hundreds. On the 25th of July the airfield is taken on the north tip of the island. The battleship *Colorado* and the destroyer *Scott* are hit by enemy eight inch guns on the island and both are out of action. The *Scott's* captain and 15 crewmen are killed. The *Colorado* suffered 25 killed and 70 wounded after taking 8 eight inch shell hits from the shore batteries. Once located, carrier planes immediately put these batteries out of action.

The landing was at 0730 hours. White Beach is small and it was not at first defended by the Japanese. They had concentrated their men and arms on "Red" and "Yellow" Beaches and expected us to land on one or both of these, but we passed on that plan. Landing on these beaches would have presented many problems for us as the surf was high and rough. Both beaches were well dug-in and defended by the Japanese. Both beaches were mined and some mines were buried in the sand. These beaches also had concrete-reinforced barriers on the beach and were underwater at part tide. Because of these things plus the way the Japanese were dug-in, it was a no go for the U.S. troops; they took the smaller but more protected beach to land on.

On White Beach the men walked ashore with hardly any opposition from the Japanese. The enemy was taken by surprise. One bad thing was the eight inch guns they had in this area and those are what hit the *Colorado* and *Scott*. Artillery from newly won Saipan covered this beach and the landings. Guns on Saipan could hit Tinian but guns on Tinian could not be trained to hit our operations on Saipan and this is one reason why Saipan was invaded first.

The Task Force (T.F. 58) off shore kept their planes airborne in case the Japanese air force showed up to try to stop the invasion. The Japanese had a number of airfields in the area and could easily hit our ships and men at Saipan and Tinian. Airfields close enough were at Truk Island, Yap, Woleai and Palau as well as Iwo

Jima, Chichi Jima and Haha Jima. Everyone was on alert for a Japanese attack, but none came. As noted, the 4th Marine Division walked ashore and behind them came all their gear, ammo, supplies, fire power, medical supplies and medics. All this is done with very little response from the enemy.

The island of Tinian is very flat and the enemy did not have many places to hide. The 2nd and 4th Marines moved forward taking the island. The Marine losses were 390 killed and 1,800 wounded. The Japanese defenders will lose close to 9,000 men, all killed. Admiral Spruance will later state that "The Tinian invasion was the best amphibious landing of the War." The Japanese had about 10,000 troops on Tinian and many of them were hit by the big guns on Saipan which were trained on Tinian, and which blasted the area prior to the invasion. It took 15 waves to get the men and equipment ashore at Tinian. Amtracs (tracked amphibious troop carriers) were used here because they could maneuver on both land and water. If the tide was out or only part way in, the amtracs could climb over the reefs and continue on to the island. Other landing craft would be stopped by the reef and could not continue in. This was the case at Tarawa, when the troops had a long walk to the inner beach area and many were killed because they were exposed in the open for too long a time. The amtracs gave the troops added protection as they were inside and not exposed to enemy small arms fire. The amtracs had 50 caliber machine guns mounted on them and they could strafe as they moved in. When the troops got ashore they could stay in the amtracs and use them like light tanks. The amtracs could carry 25 men and their equipment. They were a godsend and saved many lives. Tinian took nine days to secure. The casualty figures for the U.S. to date are 5,000 killed and missing. The Japanese losses are 48,000 men all told. The ships here in the vicinity number about 800.

The invasion of Guam on 21 July took place three days before the landings on Tinian. It had been now over a month since the Battle of the Philippine Sea and the Mariana Turkey Shoot. Guam is bombarded and bombed and strafed for 17 days on and off prior to the invasion. Guam was a U.S. possession before the war, but three days after the Pearl Harbor attack, on 10 December 1941, the Japanese invaded Guam and overran the island. Guam is hilly at its southern half and the enemy is well dug-in there, as is his artillery. There are many caves in the hills and he has built numerous bunkers and he will be difficult to get to. Big guns of the battleships and cruisers will try to hit these dug-in positions and the carrier planes will bomb and strafe them.

Guam is about 35 miles long and five to seven miles wide. It had two main towns, Agana and Agat, both on the west coast. In the bombardment of Guam, the airfield at Orote is heavily damaged. The 3rd Marine Division will have help on this invasion by the 1st Provisional Brigade and the 77th Army Division which is brought in from Pearl Harbor. The enemy was taken by surprise on Tinian, but not on Guam.

They were waiting for us and they were prepared to give stiff opposition to the U.S. troops. The American units drive forward from their respective areas and a week after landing, on 28 July, they meet on the high ground. Apra Harbor on Guam's west coast, scene of bitter fighting just a few days before, is already in use by some of our ships of the task force. The enemy had moved to the high ground area with its caves and bunkers. After a while they are defeated only by planting charges at the caves and bunkers and by the use of flame throwers to burn them out. The enemy force on Guam was estimated to be 20,000 strong. Some 9,000 remained in the hills and it took months to clean them all out. Guam was declared officially secure however, on 10 August. U.S. casualties on Guam were 3,529 killed and missing, and 1,550 wounded. 30,000 Japanese are killed or captured.

In this Marianas operation no U.S. ships other than a few landing craft were lost to enemy action. In these invasions there were 15 fast carriers and 12 escort carriers in the Fifth Fleet, and their pilots had destroyed over 1200 enemy planes and sunk over 100,000 tons of enemy shipping. The 15 fast carriers however suffered almost 360 planes lost of which 186 were lost in combat, the rest in operational accidents. A total of 241 air crewmen and six carrier sailors died in these air operations. Compared to enemy losses, American casualties were considered light.

Intrepid on 21 July, 1944 at Pearl Harbor, right, is repaired and is just about ready to depart. Enterprise (CV-6) is docked in front of Intrepid. (National Archives Photo)

Several radio controlled target planes on deck while in Pearl Harbor. (National Archives Photo)

Intrepid Returns to War
Palau and Philippine Strikes

On 10 August Air Group 18 landed on board *Intrepid* . With them came 85 aircraft and all landed without mishap. By 16 August *Intrepid* was ready to go back to war. It had been six months to the day since the torpedo hit on her stern. With two other carriers (*Enterprise* and *Independence*) and four destroyers she set a course for Eniwetok Atoll in the western Marshall Islands, with Air Group 18 on board and Captain Bolger in command. Captain Bolger celebrated the first birthday of the *Intrepid* with the following brief comment:

One hundred and forty years and six months ago today, the first United States ship to bear the name Intrepid entered the harbor at Tripoli, on the coast of North Africa, to carry out what Lord Nelson called "The most bold and daring act of the age". Manned by a crew of ten men under command of Stephen Decatur, the original Intrepid struck right into the home base of the Barbary Pirates, and under the very nuzzles of a hundred guns, that gallant crew boarded the prize vessel of our enemy and completely destroyed it.

Today, aboard this modern mighty ***Intrepid*** *we celebrate her first birthday on*

Intrepid is one year old and Commander Gaines cuts her cake. (National Archives Photo)

*this first anniversary. I know that everyone of you, my shipmates, join with me in the hope that in our further engagements with the enemy, we shall continue to be worthy of the name **Intrepid**. In cool courage and fearless bravery, in the spirit of the heroic and undaunted crews of the past, God grant that the accomplishments of our future missions may be additionally "bold and daring acts of the age".*

Intrepid's bakers had baked an enormous cake (in the shape of the carrier) for this occasion. In it went 90 dozen eggs and the cake weighed 750 pounds. Commander Gaines, *Intrepid's* Executive Officer (second in command), also delivered a message:

It is my privilege to cut the Intrepid's first anniversary cake .... she's a beauty. (This knife would be a handy size to use on some Japanese). She's a happy ship today as we realize the direction in which we are sailing. We can't win the war without smelling the smoke of battle. We are on our way again for just that. I feel permitted to extend felicitations and congratulations to the Captain on this occasion from each officer and enlisted man. We appreciate his fine message to us and I wish to assure the Captain for us all, that we are ready to do a job.

The eight day trip is uneventful and the small Task Unit (12.3.2) anchors in Eniwetok lagoon on 24 August 1944. On the 25th *Intrepid* put to sea for training exercises, returning the next day. Her steering and reduction gear problems were solved; she was whole again. On 29 August *Intrepid* reported for duty to Task Group 38.2 and got underway for combat operations. About this time the Fifth Fleet and Carrier Task Force 58 had been redesignated as the Third Fleet and Carrier Task Force 38. For a time the Japanese thought there existed in fact two different and complete Fleets. The line—up of the fleet carriers of the Task Groups (T.G.) of T.F. 38 was: 38.1 — *Hornet, Cowpens, Wasp* and *Monterey* ; 38.2 — *Bunker Hill, Intrepid, Cabot* and *Independence* ; 38.3 — *Essex, Lexington, Langley* and *Princeton* ; 38.4 — *Enterprise, Franklin, Belleau Wood* and *San Jacinto*. This last Task Group headed northwest for three days of strikes at the Bonins, then turned south to hit Yap, Ulithi and Ngulu, and then a strike at the Palaus. At the same time, the other three groups headed west to strike the Palaus. As these three left the Palaus on 9 September for strikes at Mindanao, T.G. 38.4 showed up and the Palaus got no rest. My diary beginning the 28th of August 1944 reads:

> ***28 & 29 August.*** The fleet is in and now prepares to leave to carry out the "Octagon" plan which is to bomb the island of Yap, the Palau Group, and Mindanao. I have a feeling this plan will change. I do believe Leyte will take up much of our time in the coming weeks. Our purpose now is to

destroy as much of the Japanese Air Force as we can and should we come up against the Imperial Navy, do the same to it. We leave Eniwetok on August 29th.

Intrepid is at full steam heading for the Philippines 4 September, 1944. (National Archives Photo)

30 August. *Intrepid's* second day at sea. Sea routine duties.

2 & 3 September. We reduce speed to 9 knots and we begin to refuel on the go. Some battleships and cruisers depart the fleet and head for Palau to bombard the islands.

6 & 7 September. On this day the fleet is standing out at Palau. Carrier planes strike and strafe the islands. Targets are Babelthaup, Anguar, Peleliu. We expected to find units of the Japanese fleet here, but they are no where around. No ships or planes are found here at Palau. The island of Peleliu is some 600 miles to the east of Mindanao. The island has strong defensive fortifications.

8 September. The battleships pounded the island for three days prior to our arrival. The Japanese defenders are well dug in here and have hundreds of fortified caves. The 1st Marine Division will land on the beach on 15 Sept. Meanwhile we continue our way west toward Davao, a town on Mindanao in the Philippines.

9-11 September. We head for Davao (on Mindanao) 450 miles west at full speed. We make carrier strikes against Japanese positions in the Philippines. We strike Davao and other targets on Mindanao. The Japanese fleet is not here either. They have moved again. After this strike we will head for Luzon. First we pull back for fuel 300 miles and begin refueling at 2400 hours. Continue refueling till 1200 hours. We cancel the strike on Luzon and position for a strike on Leyte and Samar. We move to Luzon late on the 14th of Sept. We are in the area for a strike on Leyte late on the 11th of Sept. We will strike Leyte tomorrow on the 12th of Sept.

12 September. On this day we strike Leyte and Samar. Leyte first. We sink one transport. We pick up the officers from the water and take them on board the *Intrepid* for interrogation. We shoot down three Japanese planes. They are Zekes. Two smaller ships are sunk. Later in the day *Intrepid* is hit while under attack. Damage is superficial. We can repair it. We down six planes (Zekes). This afternoon and into the next two days this action will continue off and on. Our first deck load of planes took off at 0800 hours. At Leyte the planes sink one small transport. One of our destroyers picks up some of the crew and they are delivered to the U.S.S. *New Jersey* for interrogation by the flag officers of Halsey's staff. Three .Zekes are set afire on the ground and two others are probables. Two small ships and sampans are also hit this morning by our carrier strike planes. General Quarters sounds at 2200 hours and again at 0525 hours. Six enemy planes are shot down at that time.

13 September. On this day the carrier *Belleau Wood* is under attack. One Zero (Zeke) is shot down. The summary for two days is: one ship sunk, 8 are left dead in the water with some burning, 6 planes shot down, and 26 probables damaged on the ground with an additional 18 more planes hit on the ground. Also hit were the usual barracks, warehouses, a large administration building and ammo dumps. All were hit hard and left afire and exploding. We have 42 prisoners total. Twenty are on the *Bunker Hill*. The *New Jersey* has several "smart ones" and *Intrepid* has some. In a Cebu harbor several ships are sunk. The dock areas and warehouses were also hit and left in ruin. Two hundred Japanese planes are destroyed in these strikes. The systematic destruction of the Japanese continues. Panay and Negros are also bombed and many areas are aflame. Many facilities on Cebu are just about totally destroyed.

During the first two days of strikes the carrier planes of the three Task Groups flew over the southern Philippine islands almost unopposed. The results of these two days were almost unbelievable: 58 enemy planes destroyed, several small cargo ships sunk, and an entire convoy of 15 cargo ships sent to the bottom by six warships of the carrier screen. The Force was running out of targets so it moved north to strike the central islands. On 12 and 13 September, again with little opposition, the carrier strikes were even more unbelievable: 478 enemy planes destroyed (173 in air combat) and at least 60 ships sunk. The 12 carriers of the Force lost nine aircraft and ten air crewmen in these four days of operations. No U.S. ship sustained as much as a paint scratch.

14 September. On this day Samar is quiet so we position for strikes on Palau again at Peleliu and Anguar. The Palaus are about 500 miles from

Edward John Ulatowski being passed over from a destroyer following his rescue. His plane went down on September 13th on a photo trip of Dinagat Island and he was rescued on the 22nd of September. Only Ulatowski, the gunner/photoman, and the pilot Ensign Daniel Weizer survived and were returned to the Intrepid. (National Archives Photo)

the Philippine area of Leyte and Samar so we get under way. Refueled on the 16th. Peleliu is under heavy bombardment by our big guns of the battlewagons and cruisers offshore. The 1st Marine Division goes ashore on 15 September. The Marines are met with heavy artillery fire and mortar fire, but they push on and into the area. During the morning 15 tons of bombs are dropped on Peleliu Island.

17 September. Standing out at Peleliu and Anguar. Carrier planes as usual bomb and strafe the islands to soften them up and knock out gun emplacements. Peleliu will be well dug in and fortified like Tarawa was and the Japanese will be dug in and hidden. The Marines are meeting stiff opposition, so the big guns move in closer and begin another bombardment of the island (Peleliu). The carrier planes cover the men all day and swoop down on troublesome positions and hit them. The main targets of ammo dumps and gun emplacements get priority in the bombing and are hit repeatedly. It took everything the troops could throw at the Japanese to burn and blow them out of their bunkers and pillboxes, and hidden caves. An area 1,000 yards north of the airfield was named "Bloody Nose Ridge" because of the action and wounded there trying to dislodge the Japanese from their pillboxes. The Japanese would not surrender they fought to the end. Every type of weapon was used here from flame throwers to mortars to tanks and hand grenades, bazookas, artillery and even hand to hand fighting with bayonets took place. Over half the Marine casualties were suffered taking "Bloody Nose Ridge" which lasted for one week. The Japanese tried to bring in more ammo and supplies and more troops from the north side of Peleliu by water but they ran into the destroyer *H.L. Edwards* which quickly made short work of disposing of the small ships and barges loaded with the supplies and ammunition. The ammo barges went sky high when hit by the destroyer fire. The taking of the Palaus cost us some 8,000 men injured, 1,200 killed or missing. The Japanese lost over 8,000 men killed and a few hundred surrendered.....

Anguar fell in five days, on 20 September. The fanatic Japanese defenders on Peleliu held out until 12 October. T.G. 38.2 and *Intrepid* withdrew from the islands on 18 September, refueled on the 19th, and headed for Luzon in the Philippines on 20 September. She arrived on station on the morning of 21 September and immediately launched strikes to targets on the main island of Luzon. The carrier attacks continued through the 22nd when they withdrew to refuel on the 23rd. My diary for this period begins on 21 September 1944:

21 September. Position is Lat. 15 degrees north, Long. 123-124 degrees east. Strike Manila and Cebu, and Negros Islands. It is 0600 hours. Action is swift and early. Carrier *Wasp* shoots done one plane (her first). *Intrepid* planes and guns down 23 planes with 2 other probables. Six out of eight torpedoes find their marks on enemy ships. Results at the moment are 6 ships sunk, with one probable, one left burning. Hits made on oil and ammo storage dumps; barracks, and administration buildings are hit. Two thousand pound bombs hit the ammo dumps. One floating drydock hit and left afire. One of our returning planes crashes on our flight deck on landing.

One other plane, a bomber, hits the barrier on landing on the flight deck. The barrier cables went up and came down on the plane cutting the plane in two. The gunner has some cuts about the face and head and a fractured elbow. The pilot is not hurt. The plane is a complete wreck and we push it over the side.

22 September. General Quarters at 0600 hours. We are under attack. Two planes shot down. Carrier *Hornet* gets one more. We secure General Quarters but are back on at 0759 hours. All ships repulsing an attack by Japanese planes. Fourteen planes are shot down and twelve others hit and turned away with a smoke trail behind them. There are four other probables. Our planes find and sink four ships with four others listed as probables. *Intrepid* loses one VF fighter in this action. The fate of the pilot is unknown and he is listed as missing. If he survived getting out of the plane, more than likely he will be spotted by one of the destroyers and picked up as the destroyers have done so many times before.

It is now 1000 hours and we are attacked again. Five more planes are shot down. The developed prints of our last strike shows ten more planes hit on the ground. We had sent up strikes at 0600 hours and 0745 hours this morning. Torpedo defense is now sounding as two groups of Japanese planes are entering our screen. We down one Japanese Judy at this time. He had made a torpedo run on us but was blown apart at a safe distance from us. We secure but within twenty minutes General Quarters sounds again. Ships repulsing another attack at this time. One of our torpedo bombers on the last strike is coming in for a landing and his wings are dipping indicating plane damage or pilot injury. We take him aboard safely. The plane is in bad shape but can be serviced and is moved down to the hanger deck. It is 0830 hours and we secure in five minutes at 0840. We went to the guns at 0600, 0745, 0759, and 0835 hours. No rest for the weary. It is now 0900 hours and planes are circling for a landing. The plane which came in all shot up will be salvaged and will be ready in about 30 hours. Another TBF (torpedo bomber—Avenger) came in for a landing with his motor aflame. The plane is also riddled with bullet holes. The plane will get a new engine and the other damage taken care of as well and be back in service in two days. We will remain here and strike Samar and other Philippine islands again.

The summary of this last action is four freighters sunk, transports hit and damaged and dead in the water and three other ships damaged. One more plane shot down. Four hours later the second strike shot down four more planes. One VF fighter is missing from the last strike. He had engine trouble and went down in the water. A destroyer patrolling our outer ring saw him ditch and picked him up and returned him to the ship.

23 September. All ships refueling. The latest summary shows that fourteen enemy planes were shot down, twelve more damaged and turned tail and departed the area. Four other planes turned and left. Four ships sunk with four other probables. Five enemy planes hit on the ground. One plane that was about to take off was hit in the vicinity of Clark Field. It was a Mike. We stop all operations at this time due to a storm which is now intensifying. The weather has been bad for two days but now we must slow down as it builds. We have been using the overcast weather quite a bit lately to cover our strikes and positions. We are not detected too easily when we move in the rain, rough seas and cloudy weather. One of the last planes to land on deck this day was a TBF and it was riddled with bullet holes and the engine was afire. The fire boys were ready. When he hit the deck, they quickly foamed the burning engine. The pilot was OK considering the shot up condition of the plane. He was very lucky.

24 September. It is 2100 hours and we are still landing planes. Today's last mission was a long one. Over 300 miles to the targets. Two of the *Hancock's* planes are on our deck. Pilots were told to land wherever they could. A third *Hancock* plane is off her landing pattern and crashes down on to our two antennas which are folded down on our starboard quarter aft. The plane flips over and is now upside down on the aerials. The nose of the plane is under the gun tub with the right wing tip resting on the first aerial support and his landing gear hooked on to the 20 mm gun tub. The pilot fell into the water and was rescued by a destroyer. The photoman-gunner walked the antenna to the flight deck. They were not injured. This was a case of the pilot not taking the wave off when he was out of position for a proper landing on deck. The plane will remain there all night. We have a bad squall up ahead, dark clouds and rain. We push on through it. Some planes still up with gas supply not gone will circle until we get through the squall area which is not too deep. We also took aboard five of the *Cabot's* planes. They were VF fighters. Some of her other planes are on other carriers as she has had a bad crack up on her deck and has an ongoing fire. Six of our VTs are missing at this time. We have had a hectic night. Our six missing planes made water landings in the dark of night. They all ditched in the same area and shared life rafts with one another. Eighteen men on rafts in water are located. One SOC (float plane) is catapulted from the battleship *New Jersey* to patrol. She found the men and the rafts. She landed on the water and taxied up to the rafts to pick up the men. Both men of the SOC get out to check the pilots in the rafts. As they do this their plane drifts away unattended. The *New Jersey* is watching them and sees their plane drift away. She heads to the area and with her hoist she lifts the plane back on board without her crew. A destroyer comes along and picks up all twenty men.

At Samar four freighters are sunk with damages to three others which are expected to sink. One Zeke plane is shot down by CAP. Our VB (dive bomber planes) and VT (torpedo bomber planes) shot down four planes. On the ground a great many planes were hit and damaged, some burning. Also four other probable planes destroyed. The air strip was heavily damaged. We lose one VF (fighter plane) over the air strip, pilot not recovered. One VF comes in for landing with the belly of the plane all shot up. These strikes were made 325 miles from the ships. The planes all have tears in the plane skin covering from the flak of the Japanese anti-aircraft guns. We can repair the torn areas of the plane skin. We have a couple of guys good at that type of repair.

25 & 26 September. Sea routine. Gunnery practice most of the day. Then withdrew from the area and started refueling as we headed toward Saipan for rearming.

Between 6 and 24 September *Intrepid,* in these actions against the Palau Islands and the central and northern Philippines, launched its Air Group 18 in 1,341 sorties, of which 1,149 were combat sorties and the rest CAP (Combat Air Patrol) and ASP (Anti-Submarine Patrol) sorties. Over 385 tons of bombs were dropped and over 280,000 rounds of aircraft ammunition (30 and 50 caliber and 20 mm) were expended. 32 enemy ships totalling 50,000 tons were confirmed sunk by Air Group 18 and another 15 ships totaling over 60,000 tons were probably sunk. 42 ships totalling approximately 35,000 tons were damaged. 36 enemy aircraft were shot down in aerial battles and 92 others were destroyed or probably destroyed on the ground during this two and a half week period. But these attacks on the enemy were not without cost. Two pilots were known killed and seven other pilots and eight enlisted men, all air crewmen, were missing and presumed dead. In spite of the casualties it was a profitable outing for *Intrepid* and its Air Group 18, and disastrous for the enemy. For the Task Force as a whole in these September attacks on all the major Philippine islands, the score was at least 840 confirmed enemy aircraft kills (probably closer to 1,000) and at least 100 ships sunk. Add to this the destruction and carnage on ground targets.

On 28 September at 0500 hours *Intrepid* and the other ships in Task Group 38.2 steamed into Saipan's Tanapag harbor. The Group stayed only long enough to replenish her bombs and torpedoes and departed on the afternoon of the 29th for Ulithi, captured without a struggle only six days earlier. Ulithi was about 480

Commander (Captain) Gaines bids farewell as Commander Kuffman and Captain Bolger bid him well 29 September, 1944. A. F. Zollo was Commander Gaines orderly at times when the Captain was not on board since August 1943. (National Archives Photo)

nautical miles southwest of the Marianas and was to become a major fleet anchorage. Task Group 38.2 sailed into Ulithi Atoll's huge lagoon on 1 October 1944. Commander Gaines, *Intrepid's* Executive Officer, was promoted to Captain in August but it took until 29 September for the dispatch to catch up to him. In my diary of this period I recorded the following:

1 October. *Intrepid* enters Ulithi lagoon. At Ulithi we load big bombs, fuel, food which took most of the day. We also got 108 bags of mail. An LST (Landing Ship, Tank) is the post office at Ulithi. It is anchored in the lagoon. We will be the flagship as Admiral Bogan comes on board with his staff. I go ashore to see the new R & R center which was set up for the crews of the ships coming and going from Ulithi. There is beer and ice cream and a rest area where the guys can just sit and shoot the breeze. Weather is overcast. Something is in the air, we are about to get a storm.....It is almost unbelievable to see what is anchored here at Ulithi. We have hundreds of ships and craft all over this huge anchorage.

2 October. General Quarters at 1105 hours. Two bogies on the screen. Secure General Quarters. It is getting dark. We have a storm coming up.

3 October. The order goes out: All ships move out to open sea. We have a typhoon coming at us. This order was given at 0525 hours early this morning. It is really getting wild out. You can't walk unless you hold on to something. The rains come and the wind increases. The ships all try to get under way at the same time. There is some confusion as they leave the lagoon. I would say we have some 400 ships in this anchorage at this time and you can imagine them all trying to get under way. All ships must turn around as the wind direction swung their sterns to face the channel. The seas are already breaking over the ships. The *Intrepid* is one of the first ships to get out to open water.

4 October. The typhoon is now in full force. We have 7 planes torn loose from their moorings on the flight deck and they go over the side into the howling seas. We carry two captain's gigs (small launches) and they both go into the sea. Some cat-walks along the flight deck edge are torn

The New Jersey and the Intrepid leave Ulithi for the Philippines as the typhoon washes waves over the bow of the New Jersey on 6 October, 1944. (National Archives Photo)

loose and hanging and swinging as the wind blows and the ship rolls. A section of the flight deck forward on the port side corner is bent and twisted up. The seas are breaking over the flight deck. Until now I have been on the flight deck and I wake up to the fact that I'm going overboard if I don't either get under cover or tie myself to something. The wind is fierce and I get down on my hands and knees and am just about able to crawl. I head for the island and my 20 mm gun. I get in the gun harness and stay there. I am soaking wet and my shoes come off. I have a life jacked on and it is wet and heavy. I doubt it would float. Men who never get sea sick are sure sick this day, me included.

After an hour or so I begin to shake; I am freezing and soaked so badly my skin is all wrinkled up. The gun is locked in position pointed straight out to sea and the wind is so strong the gun is trying to move. We had a 13 man Marine work detail as we left Ulithi and no one can find them It turned out that these men were still aboard one of the ships that had been putting supplies aboard the *Intrepid* . The carrier *Cabot* has also lost seven planes over the side.

5 October. The wind seems to be decreasing some now. we have been heading straight into the storm letting the bow take the wash of the seas. This goes on for some four to four and one half hours. We remain at sea until about 1000 hours and then turn to head back to Ulithi. As we turn the seas really push against the ship's sides and my first thought is that we are going to roll over. We had a real bad list as we turn broadside to the sea. The ship quickly came about and we were headed into Ulithi. I look like something from Mars the way my skin is all shriveled up from the rain. I am cold, wet, hungry, and tired. I do not dare go below deck just yet. I wait an hour or so, then I go down to get some dry clothes on. After it is all over I say to myself that was exciting! Some ride we had! I go topside again to sit by the gun to let my stomach settle down. Ships are entering Ulithi lagoon. They are the carriers *Hancock, Lexington, Yorktown* and with them are six small carriers. Ulithi is 110 nautical miles from Yap Island and about 900 nautical miles from Leyte at the Philippines. The wind is still some 50 miles an hour and the sea is real rough.

Where did he come from? A lot of guys ask. We have a close call — a plane just misses us as he brushed the ship. This guy got in and no one saw him. He just misses the *Intrepid* and hits the water and he blows up. It was a kamikaze plane.

6 October. On this day it is port routine. We get under way at 1600 hours. All ships departing Ulithi lagoon. Seas and winds are still heavy from the past storm. We try to refuel but seas are too rough. Part of the typhoon is still hanging around.

7 October. It is next day and still too rough to refuel Rendezvous with other three Task Groups of T.F. 38. We head for Nansei Shoto Islands (Okinawa). We again try to refuel. Some ships are done, others are not fueled. The morning is spent trying to fuel. We arrive at Okinawa on the 10th.

Okinawa and Formosa Strikes

A diversionary attack on Marcus Island is made by cruisers and destroyers on 9 October. Task Group 38.2 with *Intrepid* as its flagship, sent off its first strike to Okinawa early on the morning of 10 October. These actions north of the Philippines were designed to protect MacArthur's landings on Leyte later in the month. Airfields and aircraft were the number one targets. *Intrepid* launched over 60 fighter (Hellcat) sorties, 47 bomber (Helldiver) sorties, and 30 torpedo bomber (Avenger) sorties lasting late into the afternoon. In addition to Okinawa, several other islands were targeted, including Ie Shima, Tokuno Shima, and Amami O Shima. An Air Group 18 fighter from *Intrepid* strafed a destroyer and forced it to beach herself later in the day. Another *Intrepid* fighter downed a Tony (a single engine fighter). Airfields, hangers, barracks, work shops, a fuel dump, and two fuel barges were bombed, strafed and set afire. In addition to the beached destroyer, nine small ships were confirmed sunk and five others probably sunk. At least a dozen aircraft were destroyed on the ground with many others sustaining serious damage. Air Group 18's one-day strike was sharp, swift and destructive. For the entire Task Group, these figures are increased several times: about 100 aircraft destroyed and at least 10 ships of various tonnages sunk. There would be fewer enemy planes flying to the Philippines from Okinawa for the next couple of weeks. The Task Force lost 21 planes and nine airmen in this one-day strike on Okinawa.

From my diary of 10 October:

10 October. We are attacked. We shot down two planes. It is 0800 hours. Two more planes (Vals — single engine dive bombers) come in to the area. They are shot down at 1200 hours. We are only 375 miles from Japan at Okinawa. While our strike is taking off, two planes go in the water. They have heavy loads and as they leave the deck, they drop down and can't get altitude. The planes are one VB (Helldiver) and one VT (Avenger). One man is lost from their crews.....

The 79th sortie from Intrepid planes on Yontan San Airfield Okinawa Jima 10 October, 1944. (National Archives Photo)

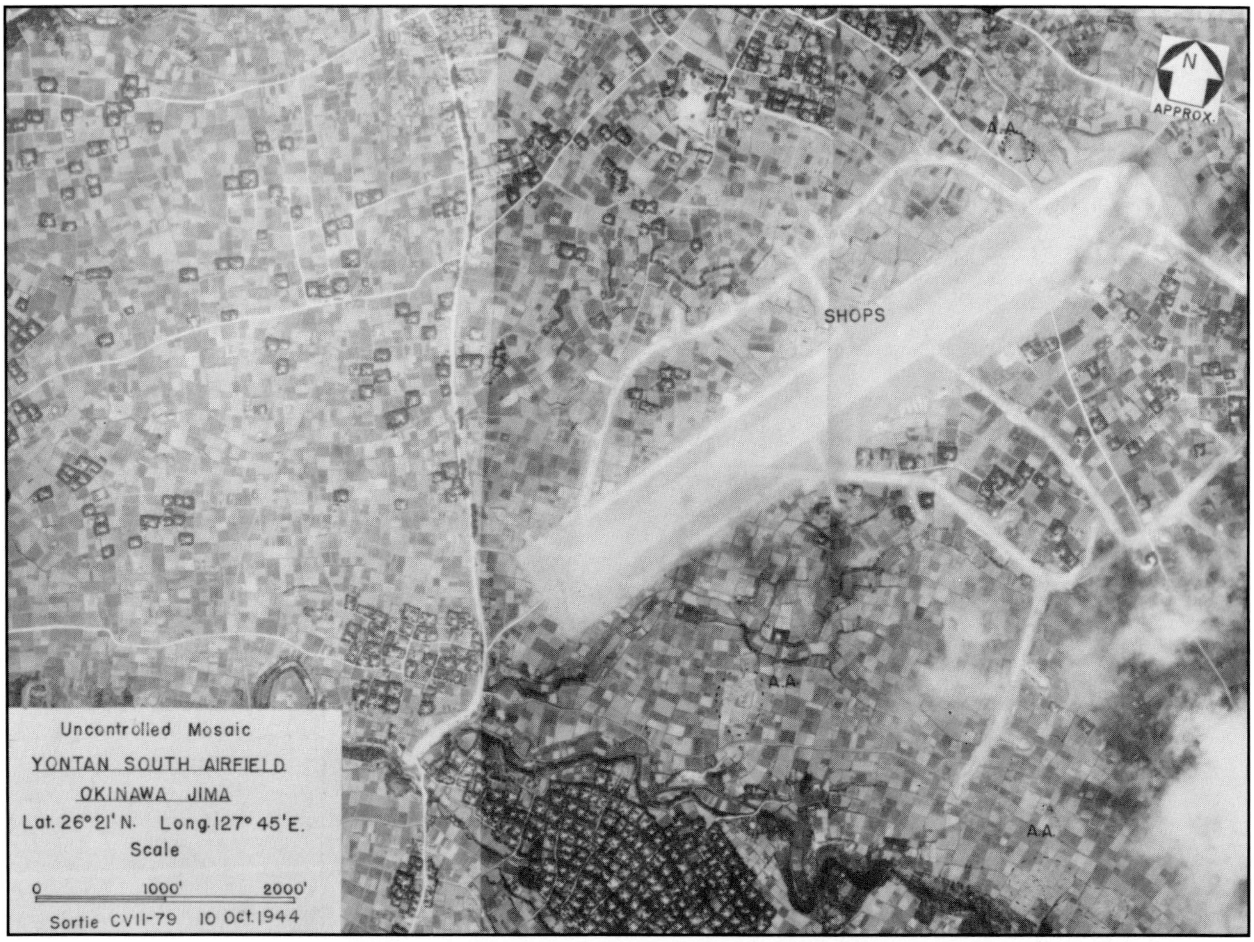

The 79th sortie from Intrepid planes on Yontan San Airfield Okinawa Jima 10 October, 1944. (National Archives Photo)

The 84th sortie from Intrepid planes on Okinawa's Tokuno Shima Airfield 10 October, 1944. (National Archives Photo)

The 84th sortie from Intrepid planes shows Okinawa's Koniya Seaplane Base at Amami O'Shima 10 October, 1944. (National Archives Photo)

Night action 12 October, 1944 with Intrepid's guns just shooting down a Japanese Torpedo Bomber. The gunners watch as the plane hits the water and bursts into flames. (National Archives Photo)

11 October. On this day we begin refueling the ships. Our position is Lat. 20°04' N, Long 125°10'E. The day is routine. No G.Q. , no T.D. (torpedo defense), no planes up, no action. Weather is better, we could not ask for a better day. We even get to eat at the right time and it is cooked food, not cold sandwiches. But tomorrow brings a change. We will be at Formosa about day break tomorrow. The strong winds that still prevail are helpful in take offs and deck landings. A full deck load of planes is ready for an early take off. They are fueled and armed. The airedales will be up early as the take off begins about 0530 hours.

12 October. The planes are up and on their way to Formosa. The planes are on a 80 mile run, one way. Being so close to so many enemy fields we expect to be attacked. Weather is and has been for days, almost perfect with some clouds in the sky which is good and bad as it can hide our planes and also hide an enemy attack force. *Intrepid* has the north end of Formosa to work over as many fields are located there.....Enemy planes begin to take to the air but before they can gain altitude our planes knock them down. Some planes are hit on the runways as they attempt to get airborne. We do much damage to the areas and the ground installations. Admiral Fukudome has his headquarters here and it is demolished. We (The Task Force) make at least 1350 sorties this day, 12 October. Massive destruction is done on the very first wave and then the second strike went in and only some 60 enemy planes came up to intercept our planes.. They are all either knocked down or hit trying to take off. On our third strike not one enemy plane is airborne. During the day several attacks are made on the ships. Betty bombers come out to hit us but our CAP gets most of them. The ones that got through are shot down by our Task Group 38.2 planes. I am told that our Task Force lost about 48 of our planes in this day's work..... It was about 1900 hours when things quieted down and we turn south. We position for another strike on southern Formosa.

A small Japanese Carrier hit off eastern Formosa 12 October, 1944 by Intrepid planes is burning.(National Archives Photo)

13 October. We begin to launch planes just after 0600 hours, about an hour before sunrise. In this area are supposed to be four airfields and we go in to hit them. As the first strike returns they report as many as 15 airfields are spotted, not four. The Task Force flies 1,000 sorties this day.....As the sun begins to go down we are attacked. Some of the planes are Betty bombers with torpedoes slung under them. Carrier *Franklin* in Task Group 38 .4 (Davison) is attacked by four Bettys. Somehow they got into the screen and radar did not pick them up. One fighter in the landing pattern shoots down one of the Bettys. A second is shot down by CAP. A third drops a torpedo at *Franklin* and flies right over the carrier. *Franklin* does some quick reverse movement and the torpedo crosses in front of the bow. The plane is then shot down. That's three of them. The fourth plane also drops a torpedo at *Franklin* which passes under her stern and the plane is now hit and he comes down on the flight deck bursting into a big ball of fire. The burning plane slid the whole length of the deck and went over the side. A fifth plane goes after the cruiser *Canberra* (13,600 tons) and drops a torpedo at her. The Canberra is hit. Canberra is in bad shape and is ordered to be towed by the light cruiser *Houston* (10,000 tons). *Houston* is hit on the 14th but manages to continue a ways further. On the 16th she is hit again and the heavy cruiser Wichita takes both *Houston* and *Canberra* in tow and all three head for Ulithi. We had planned to move away but now Halsey sets up a strike for tomorrow the 14th. He is "teed" off about all these enemy planes harassing us.

The 98th sortie from Intrepid planes 13 October, 1944. They are approaching the bombing target of the industrial area of Southeast Shinchiku Formosa. (National Archives Photo)

Intrepid's SB2Cs go after a section of the Japanese fleet 14 October, 1944. Planes 42, 43, and 46 move out to take off. (National Archives Photo)

14 October. The strike this day will be massive. The Task Force launches 146 fighters and this time we have Army help as the Air Force sends out 109 B-29s from a China base to bomb this area.

On the first day of the Formosa strikes, the 12th, all 230 of the flyable fighters on the island were sent up to ward off the carrier attacks. 80 were shot down over Formosa and another 40 as they tried to attack the carriers at sea. T.F. 38 lost 48 of its aircraft this day to combat and operational accidents.

Back to the *Canberra*, *Houston* and *Wichita*. The sea is now making up and that is because the wind has come up and it's gusty causing the tow line to snap up taut and then slack off. The *Canberra* was hit at 1845 hours on Friday the 13th and with all the action and the *Houston* being hit, the tow is not rigged and under way until about 2100 hours on the 14th of October. The tow turns to the southeast and heads to Ulithi some 1,000 miles away...... Now comes the tug *Munsee* that was with the oilers close by. She takes the tow from *Wichita* and continues on.

We now try to explain how the *Houston* was hit. On the 14th early we had sent strikes at the airfields on Formosa. That day at dusk Task Group 38.1 (McCain) came under attack by about a dozen torpedo planes. One of them put a torpedo into *Houston's* starboard side amidships. This was at 1830 hours of the 14th. *Houston* is dead in the water, all power lost. Tug *Boston* comes along and after some lost time the *Boston* takes *Houston* in tow. We now have *Munsee* towing *Canberra* and *Boston* towing *Houston.* The two tows are now making 6 knots. They travel about a mile apart and head for Ulithi. It is now decided to send a cover force with them. Halsey and Mitscher both give up ships to form a new Task Force which is called Crippled Division One. The cover force is made up of three cruisers, eight destroyers and two light carriers (*Cabot* and *Cowpens*). Then Wichita and four more destroyers join Crippled Division One and the tow.....The tug *Pawnee* now shows up and relieves the *Boston.*

There were other attacks on the Task Force and they were beaten off. No other ships were hit except the *Franklin* which had some minor damage to an area near her center elevator when an enemy plane was shot down and his wing hit the elevator as he went into the water. An attack was made on Crippled Division One, the tow, but it was driven off. In one of these attacks Task Group 38.4 (Davison) was attacked by more than 80 planes. They came from Clark Field (near Manila). About 20 of these were shot down and the rest left the area.

Intrepid and her Task Group 38.2 raided Formosa for three days, on the 12th, 13th and 14th of October 1944. 126 fighter sorties were launched from *Intrepid's* flight deck for these days, along with 87 dive bomber and

A damaged F6F plane is brought onboard on 14 October, 1944. Right wing flaps are shot to hell as is the fusalage over white star, upper point. (National Archives Photo)

44 torpedo bomber launches. Air Group 18 shot down 46 aircraft, confirmed, with seven more probables. At least 10 others were damaged on the ground. Ten small ships were sunk with one probable and numerous others damaged. Airfields, hangers, shops, warehouses, fuel dumps, docks, a drydock, a seaplane base, railroad cars and locomotives, barracks and numerous other buildings were destroyed or set on fire. Again, the damage is multiplied by other carriers in the Group.

Enemy planes harassed the Task Force all day on the 12th. And the hit and run attacks continued into the night. *Intrepid's* gunners shot down three torpedo bombers in an early evening attack. At 1840 hours about 10 enemy planes appeared on radar and the carrier went to General Quarters. At 1915 *Intrepid* gunners shot down two of them. At 2223 hours an enemy plane crossed *Intrepid's* stern and her gunners shot him into the water. *Intrepid* fueled a destroyer on the morning of the 13th.

13 and 14 October saw more of the same action as Air Group 18 searched out their targets on Formosa. *Intrepid's* radar picked up many bogeys on the 13th but these planes made no attacks.

15 October. On this day, sea routines. No Action this morning. Rig for refueling. Refueling all day. Latest report on Japanese fleet reports that they are in the Luzon area, so we head there early tomorrow after refueling is finished. We are now 500 miles from Formosa. We steer a course for Luzon.

16 October. On this day, sea routine. Yesterday morning the Task Force sent up some 400 planes, mostly fighters, to cover the B-29s and B-26s which hit Formosa. We expect to meet up finally with the Japanese fleet soon. They are reported to be in the vicinity of northern Luzon which is just about where we now are located. One bogey this morning at 32 miles and fading. Two Japanese planes shot down this morning by our combat air patrol. Torpedo Defence at 1001 hours, two enemy planes on screen. They were Judys and are shot down. We have a 4-engine flying boat nearby and believe it to be an Emily. It fades away. Secure Torpedo Defence at 1105 hours.

Meanwhile, on the 16th, a force of over 100 enemy planes from Formosa go after our Crippled Division One which has not gone too far and is still in Luzon waters. *Cabot* and *Cowpens* launch aircraft. 41 enemy are shot down. Three others head for *Houston* which has no power and only her 20 mm and some 40 mm guns operational. The first plane is low on the water, only 80 feet above the water, drops a torpedo in *Houston's* wake, and straight into her stern end. The plane is then shot down as he climbs and tries to gain altitude. Another heads for the cruiser *Sante Fe* and he is shot down.

It is questionable now whether or not *Houston* can be saved and towed as she takes on much water and settles deep in the stern. Her damage control must really be on the job as her damaged area is again sealed off and what little power she can get up continues to pump water out of her. With this second hit on her, the *Houston's* captain decided to send most of the crew to other ships just in case their condition worsens. All but those needed to keep the ship afloat are removed without a hitch or injury.

Tokyo Rose is on every day and we get other news broadcasts piped over the P.A. system. Some days she has the complete Task Force sunk. When we pull our strikes and raids on the enemy islands, she will more or less reverse the end figures. She will report only minor damages to an area when in fact we demolished it. If we have a plane lost over the target, that one or two planes become a dozen or more. When we repel enemy air attacks and perhaps down a dozen planes and receive no damages, she will report that all enemy planes returned to base after hitting and at times sinking our ships. If her figures were totally believed Task Force 38 was long ago sunk. In this case of the cruisers hit, she reports them sunk and she refers to Crippled Division One as the remains of the mighty U.S. carrier force. Each day she brings talk, smiles, and laughter to the crews with her battle reports.

In listening to her latest reports stating that Crippled Division One is the remains of T.F. 38, Halsey get an idea. He positions one group, Sherman's 38.3, west of the moving Crippled Division One and the other two groups to the east of Group .3. Sherman and Crippled Division One are now the bait to entice an attack from the enemy. This new plan gets the code name Bait Division One. Admiral Shima meanwhile is at the area where the cruisers were torpedoed. He is looking for Crippled Division One and has two heavy and one light cruiser with him and some destroyers. Carrier *Bunker Hill* of Group .2 (*Intrepid's* group) has scout planes up and they make a run on Admiral Shima's force but AA fire is heavy and they retire. Now Shima knows something is in the area, so he retreats; he doesn't want to get involved. He feels that he made the right decision to retire as now he receives a message telling him that a large carrier force is still in the area. So Shima did not take the bait. The two damaged cruisers will make it safely to Ulithi and eventually back to the States for repairs. .....

These 10 to 16 October battles around Okinawa and Formosa had cost the Japanese some 600 aircraft destroyed and about 40 ships sunk. The Task Force lost 90 planes and 64 air crewmen killed or missing. At least 600 aircraft had been sent against T.F. 38 in this seven day engagement, and this effort resulted only in putting two cruisers out of action. The enemy had, in hindsight, made a serious mistake in these Formosa battles in deciding to send from Japan virtually all available naval aircraft south to the defense of the island. Almost all were destroyed by the carrier planes and few were left to go to sea with the enemy's few remaining carriers. Enemy air strikes from now on would of necessity be land based.

Philippine Strikes and the Battle of Leyte Gulf

17 October. On this day, sea routine. Last night a fleet of enemy planes tried to hit the *Canberra* and *Reno* and *Houston* again. The Japanese fleet is no longer around this morning, we have been chasing a shadow. We have missed another chance to do battle. One of our fighters goes in the drink due to engine trouble on take off. A destroyer picks up the pilot and brings him home to us. *Intrepid's* position at 2000 hours is 19°45' N, 127°11' E.

Another pilot comes home to us from a nearby destroyer 17 October, 1944. (National Archives Photo)

18 October. On this day, sea routine. We strike Luzon today. Our targets are on the northern tip of Luzon. The results of this strike are: 2 ships sunk, 5 left burning, and direct bomb hits on them. They probably sank later. There are some 500 Japanese in the water, they are strafed by our VFs. Four enemy planes are shot down, 10 more are known to

We attack a Japanese Aircraft Carrier 18 October, 1944 and the sky is full of Japanese anti-aircraft fire. (National Archives Photo)

be destroyed on the ground and 15 more probables. The hangers and many buildings are hit by bombs and left destroyed. Tonight while landing our last strike of the day, one VF fighter off the *Hancock* goes into the water. The *Hancock* dropped flares over the water and a destroyer picks up the pilot in just four minutes time. The order to pick up the pilot to the destroyer was sent out by the *Intrepid*......In two days U.S. troops are scheduled to land on Leyte.......

19 October. On this day the *Hancock* and *Bunker Hill* are detached. We are now alone with the *Cabot* and *Independence*. Today we take on some fuel. We have a mail call. The tanker sends us mail over the wire. Ships are refueling. When we are refueled we close in for the Leyte invasion with Task Force 38.1.

20 October. On this day we are at sea routine and on this day we invade the Philippines. Our position at 2000 hours is Lat. 15°01'N, Long. 126°32'E. The Army landed on Leyte's eastern shore. Task Force 38.1 is covering the landings at Leyte while we refuel. With the assault forces at Leyte is General MacArthur; he has returned to the Philippines as he promised. A large naval force hit the eastern coast of Leyte before dawn. Then the troops went in, in two groups and split the Japanese defenders into two units. The resistance was light for a while. The troops gained control of the main coastal road and took the town of Tacloban. The price paid by the Japanese was great: 113,231 casualties. Much help was given the troops in their drive into the interior by the heavily armored "alligator" tanks.

21 October. *Intrepid's* position at 2000 hours is Lat 14°19'N, Long. 126°27'E. (about 100 nautical miles northeast of Samar). We relieve Task Group 38.1 of its duty at Leyte. We send up strikes to hit Leyte, Cebu,. Negros. The summary of this strike is: 1 Zeke shot down along with two twin engine planes, 3 other planes are shot down and one probable. Twenty planes are destroyed on the ground. Our VF fighters sank a small ship. Our VB and VT sank a ship of 150 to 300 tons. One ship of a 1,000 to 1,500 ton size was hit and is dead in the water. This ship was hit a second time and left burning and settling in the water. We have two crack ups on the flight deck. The deck is cleared and the pilots are OK. One plane is pushed over the side, the second moved to the hanger deck and will be repaired. The weather is turning bad........

22 October. On this day, sea routine. Another mail call. It is raining like hell. The cruiser *Honolulu* is torpedoed at the invasion. She will not go down and heads back for repairs. *Intrepid's* position at 2000 hours is Lat 15°35'N, Long. 128°15'E.

A three day sea and air engagement known as the Battle of Leyte Gulf, the largest naval engagement of the war, and arguably the largest in naval history, began on 23 October. It was triggered by an attempt by the Japanese to disrupt MacArthur's landings on Leyte and it almost succeeded. The complex engagement was to spell the end of the Japanese Imperial Navy as a sea power as the Marianas Turkey Shoot had done to its naval air power four months earlier. The battle saw four separate surface actions involving 280 warships as the Japanese attempted to reverse their failing fortunes. When it was all over on 26 October, the enemy had lost three battleships, four carriers (one fleet and three light, including the *Zuikaku,* the last surviving carrier of the Pearl Harbor strike), six heavy and three light cruisers and eight destroyers. The American side lost one light carrier, two escort carriers , two destroyers, a destroyer escort, a PT (Patrol, Torpedo) boat and a submarine. Both sides sustained serious damage to other ships, especially the American escort carrier fleet. One of the American escort carriers (the *St. Lô*) was sunk by Japan's newest weapon, the *kamikaze* or suicide attack by pilots who deliberately crashed their aircraft — essentially a piloted bomb — into the target. These pilots had virtually no flight training; most weren't even taught how to land their aircraft since they were not permitted to return from a mission. Other carriers were badly damaged by these attacks. I will detail these enormously important battles below for they spelled the end, after almost three years of bloody war, of Japanese naval power in the Pacific. First, my diary entries for this period show the following and serve as a brief overview of these three momentous days in October:

On 23 October, 1944 we celebrate our 7,000th landing as Ensign H. S. Measham cuts the cake, with Commander Kuffman at his side. (National Archives Photo)

23 October. On this day, sea routine. We are refueling from tankers. They have more mail for us. On this day the escort Carrier *Sargent Bay* (CVE-83) brings us out some new planes while we are at sea. We celebrate our 7,000th landing on our flight deck. Dinner is prepared and another cake is made. The landing was made by Ens. H.S. Measham .... Also on

Intrepid lands new planes brought to her from the Escort Carrier Sargent Bay (CVE-83) on 23 October, 1944. Photo taken from the Sargent Bay looking towards the Intrepid. (National Archives Photo)

At dusk on 23 October, 1944, Lt. W.B. Thompson warms up his F6F plane that will be the 1000th to catapult off our deck. (National Archives Photo)

this day the 1,000 plane is catapulted off the flight deck of the *Intrepid.* The plane is a fighter (F6F) and the pilot is Lt. W.B. Thompson. The Japanese try to stop the Leyte invasion. We split our massive fleet into four equal groups. Each will have carriers, battleships, heavy and light cruisers and destroyers. Each will have their support ships also. The Japanese split their fleet into four groups also. They call this the "Sho" plan which means "Victory" operation. The battle for Leyte continues. Meanwhile, in an action at Palawan Island, three Japanese cruisers are sunk by two of our submarines, *Darter* and *Dace.* In the next few days the *Intrepid* will participate in the following sea battles: 23 October, the Battle of Leyte Gulf; 24 October, the Battle of Sibuyan Sea; 24 and 25 October, the Battle for Surigao Straits; 25 October, the Battle off Samar and the Battle off Engaño. These battles take place one after the other.

24 October. Off the Philippines. Lat. 13°16'N, Long. 125°40'E. On this day the Japanese at this time are moving into Leyte Gulf with one of their four units. This force includes the super battleship *Musashi,* another four battleships, 12 cruisers and 15 destroyers — a lot of big guns and firepower. They should arrive at dawn of the 25th in San Bernadino Strait between Luzon and Samar. A second Japanese unit of two old battleships, one cruiser and four destroyers are at Surigao Strait, between Leyte and Mindanao. The third Japanese force of four carriers, two battleships and a number of destroyers and one cruiser are north of Leyte. The fourth Japanese force of cruisers and destroyers is close by. Also in this area is Admiral Kinkaid's 7th Fleet force with a lot of heavy fire power. In his force are 16 escort carriers commanded by the *Intrepid's* first captain, now Admiral Thomas L. Sprague. East of Luzon is Admiral Halsey's 3rd Fleet split into four groups. Each unit has four carriers, two battleships, three cruisers, and 14 destroyers. Each group also has its own support ships and each has its own commander.

Torpedo Defense went this morning at 0155 hours. We secure at 0220 hours. Another unit of the Task Force has arrived on our starboard side. They are under attack at this time. Our submarines report the Japanese fleet at the Palawan area, southwest of Mindanao. Our planes take off and are up scouting for this enemy force. We can see land on the horizon. We are 25 miles off Samar. A report comes in that our subs have torpedoed a Japanese carrier and one battleship. The latest submarine reports tell us the Japanese fleet is 210 miles from us. We go to Torpedo Defence at 0940 hours. Several bogies are on the screen. We launch our planes to go after the reported Japanese fleet position. We secure Torpedo Defence at 1130 hours. During this time we down six enemy planes; three are twin engine and are not identified Also shot down were one Jake, one Betty and another unidentified plane. This last shoot down was made by the destroyer *The Sullivans* on our starboard side. That plane was headed for us. We'll send that ship some ice cream. Another heavy cruiser is sunk by U.S. aircraft at the Sibuyan Sea battle. It was the *Chokai.*

The planes reported that the enemy force has four battleships, eight cruisers, and thirteen destroyers. One photo shows a Japanese destroyer hit and dead in the water. In the attack this morning the destroyer *Cushing* shot

Night action at Tablas Strait taken by Intrepid plane at 10,500 feet 24 October, 1944. Photo shows Japanese ships just before we attacked them. (National Archives Photo)

down a Japanese Betty who was making a run on us (*Intrepid*) on our port side. Light carrier *Cabot* takes two very near misses this morning also. One bomb fell about 500 yards astern of her. One of our VF (F6F fighter) went after that plane and shot it down. It was a Jill. CVL *Princeton* has been hit by a large bomb and she is afire now as we go into the afternoon hours. Our second strike is over the Japanese ship target, and word is sent back that they have put five torpedoes into a big battleship. Three of our planes failed to return from the strike. Three other planes got on deck OK but they are all shot up. One torpedo bomber still has his fish in the clamps as he lands and the torpedo holds, it does not shake loose. The pilot tried to drop it but it would not release. One pilot reports hitting a cruiser. AA flak was real heavy and they cannot state just how bad the enemy ships were hit. One pilot reports he dropped a 1,000 pound bomb smack in the center of a Japanese destroyer.

In this action with the Japanese fleet, we have now lost three VBs (Helldiver bombers), three VTs (Avenger torpedo bombers) and one VB made a water landing. The *Cabot* last night shot that plane down that caused us to go to Torpedo Defense at 0155 hours. They also downed a second plane soon after. Several of our planes were all shot up when they returned and one gunner of a VT was hurt real bad and so were two others gunners of our VBs. The plane's cockpit was thick with blood (the VT). It is now 1930 hours. Something is in the wind, lots of activity on the bridge and radio and radar plots. I asked a lieutenant, "What's up?" He told me we have spotted four Japanese carriers and plans are in the works to go get them and that we will head full speed to the location tonight. He says the force is off Luzon.

25 October. On this day another cruiser is sunk at the Samar area battle It was the *Chikuma* and it is sunk also by U.S. aircraft. Also sunk at Samar on the 25th was the cruiser *Suzuya* by our planes. Also on the 25th of October at Mindanao the cruiser *Mogami* was sunk by our planes. And on this day, at the Battle of Surigao Strait, the battleships *Yamashiro* and *Fuso* were sunk in the strait by U.S.aircraft. Again on the 25th the battleship *Musashi* is sunk in the Sibuyan Sea battle. *Intrepid* planes are involved in all these sinkings and we take credit for the *Yamashiro* sinking. Japanese ships sunk in three days at these locations are: 23 October, the three heavy cruisers sunk at Palawan Island, *Maya, Atago, Takao;* on 24 October, the heavy cruiser *Chokai* is sunk in the Sibuyan Sea; on 25 October, the two cruisers *Chikuma* and the *Suzuya* are sunk at Samar, and the cruiser *Mogami* is sunk at Mindanao. On this same day at Surigao Strait the battleships *Yamashiro* and *Fuso* are sunk. At the Sibuyan Sea on this same day the battleship *Musashi* is sunk. Also on the 25th in the Philippine Sea are four carriers and they are sunk. They are the *Chitose, Chiyuoda, Zuiho* and the *Zuikaku.* The super battleship *Yamato* is damaged also on this day and she retires and heads for repairs. At least eight destroyers are also sunk in these three days.

Now it is necessary that we return to the beginnings of these momentous events and examine in some detail this enormous American victory at sea. It took only three days in October of 1944 but when it was over the sea lanes to the Japanese home islands had been cleared and the result was the eventual capitulation of the enemy. By 26 October, the United States Navy would own the Pacific Ocean and would go and come as it pleased.

First, the forces involved. The main U.S. force is Carrier Task Force 38. It is composed of four Task Groups of four carriers each, and each with its screens of battleships, cruisers, destroyers, destroyer escorts, and with support ships. In addition to this powerful force the U.S. had the Seventh Fleet in Philippine waters consisting in part of over 400 transports, cargo ships, landing craft, six older battleships with cruisers, destroyers, destroyer escorts, and sub chasers which was the Leyte landing force. With this enormous landing force was a carrier group (T.G. 77.4) composed of 18 escort carriers, nine destroyers and 14 destroyer escorts. This carrier Task Group was divided into three groups and known by their radio code names as Taffy One, Taffy Two and Taffy Three. T.G. 77.4 was commanded by Admiral Thomas L. Sprague, *Intrepid's* first Captain.

The Japanese have four powerful groups underway to the Philippines, but only one has carriers.

(1) This four-carrier northern force is headed by Admiral Ozawa and its southern course from Japan will take it well off Cape Engaño at the north tip of Luzon. It will be the bait to draw off U.S. fire power from the Leyte area. Ozawa's force with barely over a hundred planes on the four carriers, is expendable. In this group are two old battleships fitted with short flight decks, three cruisers and ten destroyers.

(2) The Japanese central force, commanded by Admiral Kurita, originates in the Singapore area and sailed to Brunei Bay off Borneo where it refueled. It then turned north and then east to sail south of Mindoro into the Sibuyan Sea and thence into the San Bernardino Strait between Luzon and Samar. In this group are five battleships, ten heavy and two light cruisers and fifteen destroyers.

(3) The Japanese southern force broke off from Kurita's group and proceeded east across the Sulu Sea, into the Mindanao Sea and then to the Surigao Strait between Leyte and Mindanao. It is commanded by Admiral Nishimura. His force

We wait at our general quarters on 25 October, 1944, as Japanese planes are on the way to us. (National Archives Photo)

consists of two battleships, four cruisers, and eight destroyers. The big guns of U.S. Admiral Oldendorf's battleships and cruisers were at this Strait and Taffy Two was just off shore.

(4) The fourth Japanese unit of three cruisers and four destroyers, commanded by Admiral Shima, is headed south from the west side of Formosa. This unit planned to join Kurita and Nishimura at Leyte but would get only to the southern tip of Negros Island in the heat of the battle to come.

Prior to the landing on Leyte Gulf early on 20 October; the mission of Admiral Tom Sprague and Admiral Oldendorf was to knock out airfields on Cebu, Panay, Negros, and fields at northern Mindanao. Admiral Sprague had the 18 escort carriers under him (the Taffys) positioned in and around the Leyte area; Taffy One was south off Mindanao, Taffy Two in the center was off Leyte and Taffy Three was north off Samar. On the morning of the landings four of these small carriers were to cover the invasion at Leyte. Oldendorf's battle group was the heavy fire power and he bombarded the area and hit any shipping found headed to or away from Leyte. With the bombardment finished late on 19 October, Oldendorf's fire team withdrew for the invasion.

On 20 October a U.S. ship was an early casualty in the impending sea battles. The cruiser *Honolulu* is about five miles off shore and Admiral Ainsworth's flag is in *Honolulu*; he has just completed bombardment of the invasion areas at Leyte and was standing by for a call to send more fire onto the beaches. The ship's speakers come alive with a shout by lookouts, "Torpedo dropped port quarter!" The captain looking to port saw the enemy torpedo plane and the torpedo wake coming to him on a line at port beam. Engines of *Honolulu* had been at one-third speed; the captain orders full speed astern and at the same time swung the ship's bow right. Hopefully the torpedo would parallel the ship's port side. There was not sufficient time to do this and the torpedo hit *Honolulu* port side forward of the superstructure. Had the ship not moved as she did the torpedo likely would have gotten her in the after magazine and certainly the after engine room.

Honolulu had a ripped open port side 25 by 29 feet. 60 men of the crew were killed including some officers and part of the damaged area was the Marines' compartment. The time was 1600 hours and *Honolulu* was dead in the water and took on a list of some 15 degrees. Everything that should have been done at the moment the torpedo hit, was done and done right as some fifteen minutes later the shut down engines roared to life and *Honolulu* slowly moved nearer the shore area and at the same time transferred her fuel from one side to the other, correcting the list. Destroyers come along side and remove her wounded and assist in pumping the water out of her.

At about 1700 hours the cruiser is taken in tow by fleet tugs and heads for repairs. But her day is not yet over as enemy planes appear in the area beginning at

1800 hours. Smoke screens are let out by the destroyers. The alert continues until 2030 hours. During this time two U.S. ships are hit by friendly fire due to the smoke screen. *Honolulu* is one which is hit by this U.S. fire. She takes 20 mm fire on one of her 40 mm gun mounts which explodes some ammunition and five crew members are killed and eleven wounded in that 40 mm battery.

While Task Force 38 was scouring the Philippines for targets, and finding many, two U.S. submarines near Palawan Island, the *Darter* and the *Dace* are operating as a team and on the surface on the early morning darkness of 23 October. They would draw first blood in the great battles about to start. At 0130 hours *Darter* has a radar contact. The indication is at least a dozen ships. Both subs up speed to close the distance and get a look at what lies ahead. *Darter* and *Dace* have stumbled upon Kurita's central force. Kurita's position has been a mystery to Admiral Halsey. Task Force 38 knew of its existence but not its location or direction. It now is an hour before daylight and *Darter* sends a message which Halsey receives at 0620 hours. The message gives position, direction, and number of ships in the enemy force. Kurita is making 16 knots and the weather is clear. Kurita does not have pickets ahead of him to protect the column. Prior to sending the message to the fleet, *Darter* and *Dace* take positions to make torpedo runs on the heavy force in front of them.

The enemy ships are in two columns coming right at the two U.S. subs. Both subs dive and *Dace* moves to the right column, while *Darter* positions herself outboard of the left column. Both ships wait until the ships are at a range of about 1,000 yards; it is 0630 hours. *Darter* sends out her six bow torpedoes at the heavy cruiser *Atago. Darter* then turns quickly and sends her stern torpedoes at the second cruiser in the column which is the *Takao.* Elapsed torpedo run time is about four minutes. The bow-fired torpedoes go to the *Atago* the four stern torpedoes go to the *Takao. Darter* goes deep but before she does she takes a look at both targets. *Atago* is so close that she is almost over *Darter. Atago* is a mass of fire and is sinking by the bow. *Takao* is hit under her stern by the first two torpedoes and two others are about to hit her. *Darter* dives deep as more explosions are heard. By now enemy destroyers are dropping depth charges over *Darter* who does not move away and so the destroyer's depth charges are all wide outside of *Darter's* position. Loud noises are now heard and *Darter* begins to shake. Could it be that the *Atago* is coming down on top of *Darter*? *Darter* moves away as the destroyers also moved out. It is 0655 hours and the *Atago* goes down. Many of her crew go into the water including Admiral Kurita. The *Takao* is also sinking.

On the other side of the column *Dace* had fired four torpedoes at the heavy cruiser *Maya,* the third ship in the column. Again *Dace* like the *Darter* begins to shake violently and all kinds of noise is heard. She too dives deep as the *Maya* exploded and all traces of her disappeared. Now *Dace* has many depth charges dropped over her with some coming very close but she is not hit.

Takao does not sink however and is taken in tow westward, screened by two destroyers. These three are out of the fight. After dark the two subs surface and decide to follow the enemy force and perhaps finish off the *Takao*. They later break away from Kurita's ships and at 0100 hours of 24 October, *Darter* has the bad luck to run aground on a coral reef. *Dace* takes off her crew, partially destroys the *Darter* and heads for Australia. Beside the sinking of the two cruisers and further reducing Kurita's force by the three ships that withdrew (later sunk), the great importance of this engagement is that T.F. 38 now knows where the enemy central force is, and Halsey gears up for the attack.

At this time, 23 October, one Task Group (38.1) was on the way to Ulithi for replenishment. It would soon be recalled. At dawn on the 24th the remaining three groups launched search planes toward Kurita's last reported position. A plane from *Intrepid* sighted Kurita's 27 remaining warships at about 0815 near Mindoro. Almost immediately Nishimura's seven ship southern force was sighted by *Enterprise* planes in the Sulu Sea near Mindanao to the south. Halsey ordered T.G. 38.1 to reverse course, refuel, and join the main force. Then he ordered all carriers to attack. But the enemy beat him to the first punch. Before the U.S. carriers could get off their main strikes the Japanese sent out 50-60 plane waves to strike the Task Force, concentrating on 38.3, the most northern Task Group. Seven fighter pilots from *Essex* shot down half of the first wave. But at 0930 a Judy broke through the screen and the cost would be high. From my notes for this day, 24 October 1944:

> *Intrepid,* flagship of Task Group 38.2 with Admiral Bogan in *Intrepid,* is positioned off the San Bernardino Strait area.
>
> *Intrepid's* Task Group is enormously powerful, and is made up of the following ships and men: In overall command of 38.2 is Rear Admiral Gerald F. Bogan. His flag is in *Intrepid.*
>
> **Carrier *Intrepid*, CV-11, (Flagship) Captain J.F. Bolger.**
> Air Group 18, Commander W.E. Ellis
>
> **Carrier *Hancock*, CV-9, Captain F.C. Dickey.**
> Air Group 7, Commander J.D. Lamade
>
> **Carrier *Bunker Hill*, CV-17, Captain M.R. Greer**
> Air Group 8, Lt. Comander R.L. Shifley
>
> **Carrier *Cabot*, CVL-28, Captain S.J. Michael**
> Air Group 29, Lt. Commander W.E. Eder

Carrier *Independence,* CVL-22, Captain E.C. Ewen
Air Group 41, Commander T.F. Caldwell

Battleship Division 7, Rear Admiral Oscar C. Badger
Battleship *Iowa,* Captain A.R. McCain
Battleship *New Jersey,* Captain C.F. Holden

Cruiser Division 14, Rear Admiral F.E.M. Whiting
Light Cruiser *Vincennes,* Captain A.D. Brown
Light Cruiser *Miami,* Captain J.G. Crawford
Light Cruiser *San Diego,* Captain W.E. Mullan
Light Cruiser *Oakland,* Captain K.S. Reed

Destroyer Squadron 52, Captain J.P. Wombly, Commanding
Destroyer *Owen,* Commander C.B. Jones
Destroyer *Miller,* Lt. Commander D.L. Johnson
Destroyer *The Sullivans,* Commander R.J. Baum
Destroyer *Tingey,* Commander J.O. Miner
Destroyer *Stephen Potter,* Commander L.W. Pancoast

Destroyer Division 104, Captain W.T. Kenney, Commanding
Destroyer *Hickox,* Lt. Commander J.H. Wesson
Destroyer *Hunt,* Commander H.A. Knoertzer
Destroyer *Lewis Hancock,* Commander W.M. Searles
Destroyer *Marshall,* Commander J..D. McKinney

Destroyer Squadron 50, Captain H.B. Jarrett, Commanding
Destroyer *Halsey Powell,* Commander S.B.D. Merill
Destroyer *Cushing,* Commander L.F. Volk
Destroyer *Colahan,* Commander D.T. Wilber
Destroyer *Uhlmann,* Commander S.G. Hooper
Destroyer *Benham,* Commander F.S. Keeler

Destroyer Division 106, Captain B.E. Tompkins, Commanding
Destroyer *Stockham,* Commander E.P. Holmes
Destroyer *Weiderburn,* Commander C.H. Kendall
Destroyer *Twining,* Commander E.K. Wakefield
Destroyer *Yarnall,* Commander J.H. Hogg

We catapult planes off the deck headed for a Kamikaze base to knock out planes on the ground. Two others wait to be catapulted and will be followed by a full deck load of planes. Admiral Bogan is in command. (National Archives Photo)

Five carriers (capable of launching at least 350 aircraft), two battleships, four cruisers and 18 destroyers. The make-up of Task Force 38 was four times this strength.

We were there in anticipation of making contact with and destroying Admiral Kurita's central force. We know they were about to enter the Strait and we would wallop him in the Strait with a strike by our planes and also when he came out of the Strait. *Intrepid* and Task Group 38.2 had the position to 'do in' Kurita. Admiral Sherman's Task Group 38.3 was north of T.G. 38.2, while Admiral Davison in T.G. 38.4 was positioned to the south off Samar about 60 miles. We are about to begin the first of many sea battles off the Philippines and the first will be with Admiral Kurita's central force. This action will take place in the Sibuyan Sea before he is able to exit the San Bernardino Strait. Task Group 38.1 under Admiral McCain, was on its way to Ulithi at this time to rearm and re-provision. Admiral Halsey ordered McCain to reverse course and return to the area......

October 24 dawned fair and was favorable for our plane operations that day. All ships have launched search planes at dawn, each to a different area. They are looking for the reported Japanese ship sightings of which we now have at least three reports of three different ship groups. The carriers send search planes to the west side of the islands (across land areas). One group searches the west coast of Luzon and continues into the Sibuyan Sea Area and flies over the Mindanao Sea and Sulu Sea areas. The enemy force in this area is Admiral Kurita's central force, supposedly heading through the Sibuyan Sea to the San Bernardino Strait. His flag is on the super battleship *Yamato*, sister ship to the *Musashi*, which is with her at this time. At stake in these actions is Leyte and, eventually, the Philippine Islands. Who is going to control them?

At about 0815 hours of October 24 an *Intrepid* plane sighted Kurita's central force and within minutes the plane sends a message to Halsey: "Central force located in Tablas Strait — appears headed for San Bernardino Strait." The Tablas Strait is just off the southeastern coast of Mindoro. Halsey had issued orders that no ship enter the Strait as it was mined and the shoal areas were known only to the Japanese. No need to enter the Straits anyway, as the carrier *Intrepid* is only some 50 miles off shore and the planes would have no trouble hitting Kurita's force. Halsey now orders Sherman and Davison to close on the *Intrepid* and all carriers to strike Kurita. All three fast carrier Groups, 2, 3 and 4, are lined up to hit Kurita. The die is cast. The planes are ready, armed and fueled and they are launched by mid morning off the carriers headed to Tablas Strait. We will return to detail these strikes shortly, but first the action at and over the Task Force at sea.

Admiral Kurita had spotted the *Intrepid* scout plane at about 0810 hours that morning and so he expected something as he sat in his flag ship, the mighty *Yamato*. As this strike begins and just before the carrier planes arrive over Kurita's force, a group of Japanese planes left a field on Luzon and headed for their nearest target which was Admiral Sherman's Task Group 38.3, further north than Intrepid (T.G. 38.2). The first wave of enemy planes over Sherman's group was at least 50 in number. Three waves come at Sherman's group, the second and third wave each having some 60 planes. Sherman's planes are in the air to intercept. The weather has turned to rain and it is now overcast. The enemy planes were met and shot down or turned tail. Commander David McCampbell is up with only some seven planes of his air group, which is based on the carrier *Essex*. He and his men climb high and come down on the attacking Japanese planes and they have a field day, hitting or dispersing the enemy in all directions. Commander McCampbell had nine shoot downs, his wing man had six and the other planes had at least two each. Only 18 planes of that group got away. Air Group 32 off the *Langley* was also involved in the action over the fleet with one of the enemy waves attacking Sherman's 38.3. Involved also in these three enemy raids on Task Group 38.3 were planes from the light carrier *Princeton* (CVL-23).

After it seemed that the air action was over, the planes began to return to their carriers because some were low on fuel as some of the planes had been airborne for some one and a half hours or more. One enemy plane came out of nowhere, a Judy bomber dropping down from very high where he had been waiting, and dropped two 500 pound bombs through the flight deck of the *Princeton*. The time is about 0940. The bombs went through both the flight and hanger decks and exploded on the next deck below in the area of the ship's bakery. The hanger deck was a raging inferno and parked planes began to blow up. So great was one of the many explosions, that No. 2 elevator was blown into the sky some 100 feet. When it and all parts of it came back down, it landed on the aft elevator, putting it out of commission also. In about 30 minutes the captain orders all hands to abandon ship except for those in damage and fire control.

The weather had now partially cleared and other ships tried to get along side the burning *Princeton* to remove her crew. Because of the flight deck overhang and some gun mounts below the flight deck, the destroyers could not get close enough to the men on the *Princeton's* forward end. This is when the *Princeton's* crew began to go over the side on ropes, some jumping into the water. The sea is now running high and some of the men drift aft of the ship. The destroyers have nets over the side and the survivors from the *Princeton* climb on board with the help of the ships' crews. On hand now are the cruisers *Reno* and *Birmingham* who are about to help fight the fires on the *Princeton*. *Birmingham* shot lines over to the men on the *Princeton* and tied to those lines were fire hoses from the *Birmingham*. Also onto the *Princeton* went the fire crews of the *Birmingham*, which numbered some 38 men in the first party. They and the *Princeton's* fire crews tackled the fires. Some two hours later it looked like the men would have the fires under control.

At least 400 men have now been picked up out of the water by several destroyers. At 1215 hours there is a radar report of incoming planes. It is now about 1330 hours and minutes later a contact is made on an enemy sub in the area. The *Birmingham* quickly lets go from the *Princeton.* The enemy attack never materialized as some 50 miles to the north carriers *Lexington, Langley* and *Essex* took care of it, as well as another large enemy group that came along. It was said that these planes came from Admiral Ozawa's four carrier force to the north and that those planes that got away from our carrier planes did not go back to their carriers. Instead, they landed at a field on Luzon. Some did in fact fly from Ozawa's carriers to the north, but most came from Luzon. With these threats over, the rescue began again on the *Princeton.* All fires are about out on *Princeton* except the one aft which is now threatening the torpedo storage area.

At 1445 hours the *Birmingham* came back along side the *Princeton* to continue its fire fighting duties and then it happened. At 1525 hours a massive explosion on the *Princeton* sent hot steel and fire over onto the *Birmingham.* The explosion is so great that the stern and half the flight deck of the carrier *Princeton* are gone. Some of *Birmingham's* crew, all top side helping out, were killed instantly. Men had legs and arms severed from their bodies. Blood was every where, a horrible sight to see. *Birmingham* received slight damage but it was sufficient to send her back for repairs.

The men on the *Princeton* that survived the blast are removed and the *Princeton's* gutted hull is ordered sunk. A destroyer and a cruiser fire eight torpedoes at her and she is blown up and sinks. 106 *Princeton* men were killed and another 190 wounded, of her 1500 man crew. The *Birmingham* suffered more. 230 of her men died instantly in the explosion and many of her 420 wounded died later. These numbers added up to half her crew. The *Princeton* was the first U.S. carrier to be lost since the *Hornet* went down in the Solomons two years before.

If this was a bad 24 October for the *Princeton* and *Birmingham*, it was worse for Admiral Kurita and his central force off Mindoro. *Intrepid* was the first to launch planes against Kurita's force when its position became known. 48 planes went out in the first strike. It is just after 0900 hours and *Cabot* then launches her planes. These two waves hit Kurita's ships at about 1030 hours. In the meantime a second strike force is readied back at the carriers and about 46 planes from *Intrepid* and the same number from *Cabot* take off on the second strike against Kurita. This second strike takes off at 1045, forms up, and reaches Kurita's force at 1245 hours as the first strike turns away to return to the carriers. A third strike force takes off at 1345 hours and is headed to Tablas Strait and Kurita's fleet. In the meantime carriers *Lexington* and *Essex* launch planes toward Kurita also. Now Admiral Davison's planes in T.G. 38.4 take off at 1315 hours and they head for Kurita. On the first strike the planes also located Admiral Nishimura's force which was off Negros Island.

In the first strike on Kurita *Intrepid* planes made a bomb hit and a torpedo hit on the super battleship *Musashi.* On the second strike *Intrepid* planes and *Cabot* planes hit *Musashi* again. This time four torpedoes hit her. On the third strike *Intrepid* planes again go after the *Musashi* and again make several hits on her. The pilots talking to each other over Kurita's force is piped over our PA system and they come in loud and clear. It is now 1500 hours and our planes return to Kurita's force and they are intent on sinking some enemy ships. *Musashi* is now slowed and almost dead in the water and has fallen away from the main force. The planes of *Franklin* and *Enterprise* are returning to their carriers as *Intrepid* planes arrive with *Essex* and some of the *Cabot* planes over the enemy ships. The *Musashi* is hit with ten more torpedoes and more bomb hits are made on her. As of now the *Musashi* has taken 19 torpedoes and at least 17 bomb hits. The pilots are talking of this and we can hear them plain as day. The *Musashi* is now at a list of some 25 degrees and taking on water very fast. It is only a matter of time now before she goes down. The pilots are talking about this and how the *Intrepid* will take credit for her sinking.

It is about 1800 hours now as our planes turn to come back to the carriers. Several planes which arrived late remain and they witness the *Musashi's* end at about 1930 hours. They reported that her list was perhaps over 30 degrees and that she sank bow first and the last picture of her was her stern sticking straight up in the air, her rudder and screws showing as the sun behind her began to settle also. Later it will be reported that the *Musashi* carried a crew of 2290 men, 112 of them officers, and that 985 men and 40 officers were killed. Survivors were picked up by the enemy's 13 destroyers in the area. The battleship *Haruna* was hit and damaged as was the *Nagato* which was hit twice. The sister ship to the *Musashi*, the *Yamato* is also hit twice. The cruiser *Myoko* has her props and shafts damaged by a torpedo and she is out of action. The *Haruna, Nagato* and *Yamato* remain on line as their damage is not sufficient for them to head for repairs.

Kurita's force is still powerful, even though the *Musashi* is sunk and he has lost the three cruisers at Palawan sunk or damaged by our subs and a fourth cruiser in the action at Tablas. He still commands four battleships, eight cruisers and thirteen destroyers. Kurita now has pulled back some in order to make certain repairs to the slightly damaged *Yamato* and the *Haurna* and *Nagato.* This action took place at the entrance to Tablas Strait, and Kurita is now some 20 miles west of there. Our planes got back to *Intrepid* just at sunset on this 24th of October. ASP and CAP planes remain in the air for a time. With all the flights made against Kurita's force and all the planes which took part, the airedales and handlers are now real busy refueling and rearming the planes and getting them spotted on the flight deck for the next strike. The pilots have much to talk about and the sign men do not wait for daylight as they bring our score up to date. Task Force 38 lost 20 planes (18 in combat) in these Tablas strikes, ten more in air defence of the fleet and in search sweeps, and 26 when the *Princeton* was hit.

Kurita never had air protection over him except his own small CAP made up of the cruiser and battleship float planes which our planes knocked down as soon as they got over Kurita on the first strike. With Kurita still in the Sulu Sea we will again go after him soon for he turns to head for San Bernardino Straits again. Search planes report his course and speed. And this brings us to the action of Admiral Tom Sprague's Taffy units 1, 2, and 3 at Leyte and off Samar. Kurita's and Nishimura's forces were to meet at Leyte; this never took place as we hit Kurita at Tablas Strait and reversed his course.

Nishimura's small southern force of two battleships, one heavy cruiser and four destroyers would next come under fire. He had been discovered on the afternoon of the 24th and attacked; some hits were made on his force and then this attack was called off in order to concentrate all striking power on Kurita to the north. Nishimura continued his course toward Leyte Gulf while Kurita was under attack. By the early morning of 25 October Nishimura was about to enter Leyte Gulf from Surigao Strait. U.S. fleet scout planes were following his course and we were waiting for him.

Admiral Kinkaid commands the Seventh Fleet which included the escort carrier groups, the Taffys, and he set an ambush for Nishimura as the Japanese force entered the 12 mile wide Surigao Strait leading to Leyte Gulf. Kinkaid deployed some 40 motor torpedo boats, six old battleships, four heavy and five light cruisers and 28 destroyers, and waited for Nishimura to show up. Behind Admiral Nishimura's group is a force commanded by Admiral Shima with two heavy cruisers, a light cruiser and four destroyers. Kinkaid was not worried about Admiral Shima; he would take on Nishimura first. Kinkaid's first order went to Admiral Oldendorf who commanded the "big gun" ships — battleships, cruisers and destroyers — to take up a position at the Leyte Gulf entrance to Surigao Strait. Under Oldendorf's immediate command were three heavy cruisers, one of which is his flagship, the *Louisville*, three light cruisers and nine destroyers. These ships are called the 'left flank.' The 'center force' of this battleline is made up of six battleships (five of them raised from the bottom of Pearl Harbor), with Admiral Weyler's flag on *Mississippi*, and six destroyers. The 'right flank' is in command of Admiral Berkey. His flag is on the light cruiser *Phoenix* with the cruiser *Boise* also there. The heavy cruiser H.M.A.S. *Shropshire* (Australian) is also part of the right flank as are six more destroyers. A roaming screen of seven destroyers circles these forces. The P.T. boats were plying the waters just south of Leyte and west of Mindanao in the Mindanao Sea. There are 28 troop transports in the Gulf protected by other cruisers and destroyers. General MacArthur is on the cruiser *Nashville*. All this was in the late night blackness of the 24th of October.

Pilot Lieutenant N.E. Johnson and gunner AM 3/C R.G. Masterson just got back from a strike on a Japanese fleet 25 October, 1944. Note the camera that the gunner carries. (National Archives Photo)

The seas are calm, the moon is up but then it became overcast and the night became pitch black; then it rained. The night is quiet, nothing is moving to light up the radar screens. What happened long ago at Savo Island, Empress Augusta Bay, the "Slot" and "Iron Bottom" Bay at the Solomons when we were surprised and lost some ships, would not happen here. We were ready for what was to come. The P.T.'s, the destroyers and the "battleline" of big guns were all in position and waiting.

Admiral Nishimura's southern force was to arrive off Tacloban in Leyte Gulf at about dawn or just after 0430 hours. Kurita's force was also to arrive at the same time and the two enemy forces were to hit the U.S. amphibious forces in Leyte Gulf from two sides. Kurita had already tangled with the fleet carrier planes earlier on the 24th and that action did not end until about 1600 hours that day. At that time Kurita was in contact with Admiral Nishimura to tell him he had been delayed by the seven hour battle with the U.S. carrier planes.

Admiral Nishimura now sends the cruiser *Mogami* and three destroyers ahead of him into Surigao Strait to patrol the waters. These ships now come into the

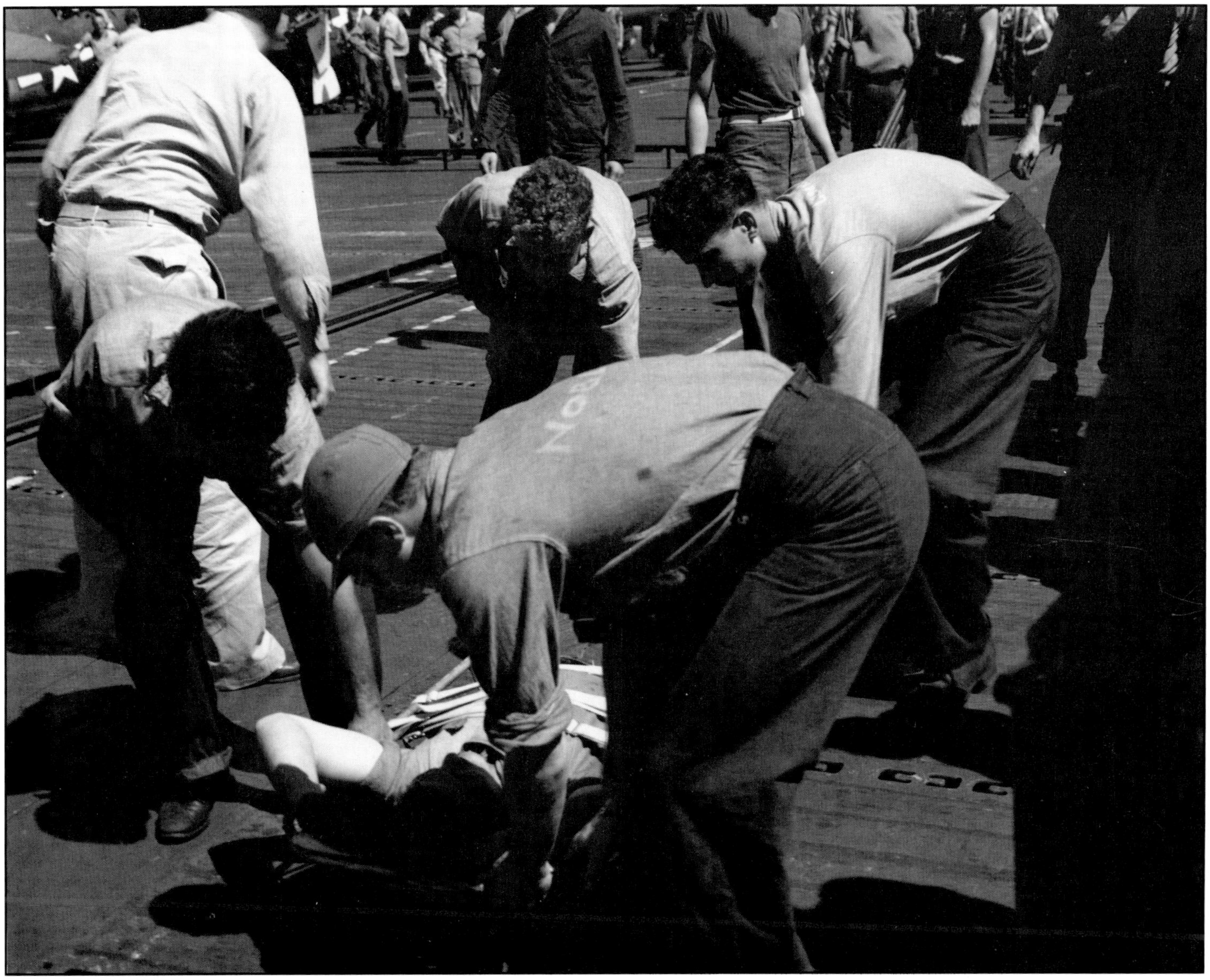

A pilot, although injured, lands on our deck safely after participating in raid on Japanese fleet 25 October, 1944. He is taken to Sick Bay for treatment and will be back to fly another day. (National Archives Photo)

area of the waiting P.T. boats and the battle is joined. The little P.T. boats attack and fire torpedoes at this four-ship group, and miss everything. Two of the small boats are damaged by destroyer fire, and the torpedo boats retire behind a smoke screen. Oldendorf continues to wait patiently as Nishimura approaches, totally unaware of what lies in waiting for him. At about 2245 hours of the 24th the first group of six U.S. destroyers move into the Strait; radar shows Nishimura's force at 4 1/2 miles range. The range decreases and the destroyers fire off 27 torpedoes at Nishimura's seven ship force, and are in turn fired upon. None of the U.S. destroyers are hit. It was a long seven minute torpedo run and several explosions are heard as the U.S. ships pull back out of Nishimura's gun range. The battleship *Fuso* was hit in this first attack but not the battleship *Yamashiro*. Now a second group of destroyers moves into position for another torpedo launch, and they get more hits. This time three destroyers are hit and one sinks almost immediately. A second goes down a few minutes later. The third destroyer has a gaping hole in her bow and turns to leave the battle area. These first two attacks fired off 47 torpedoes. Now a fierce torpedo

Philippine Strikes and theBattle of Leyte Gulf

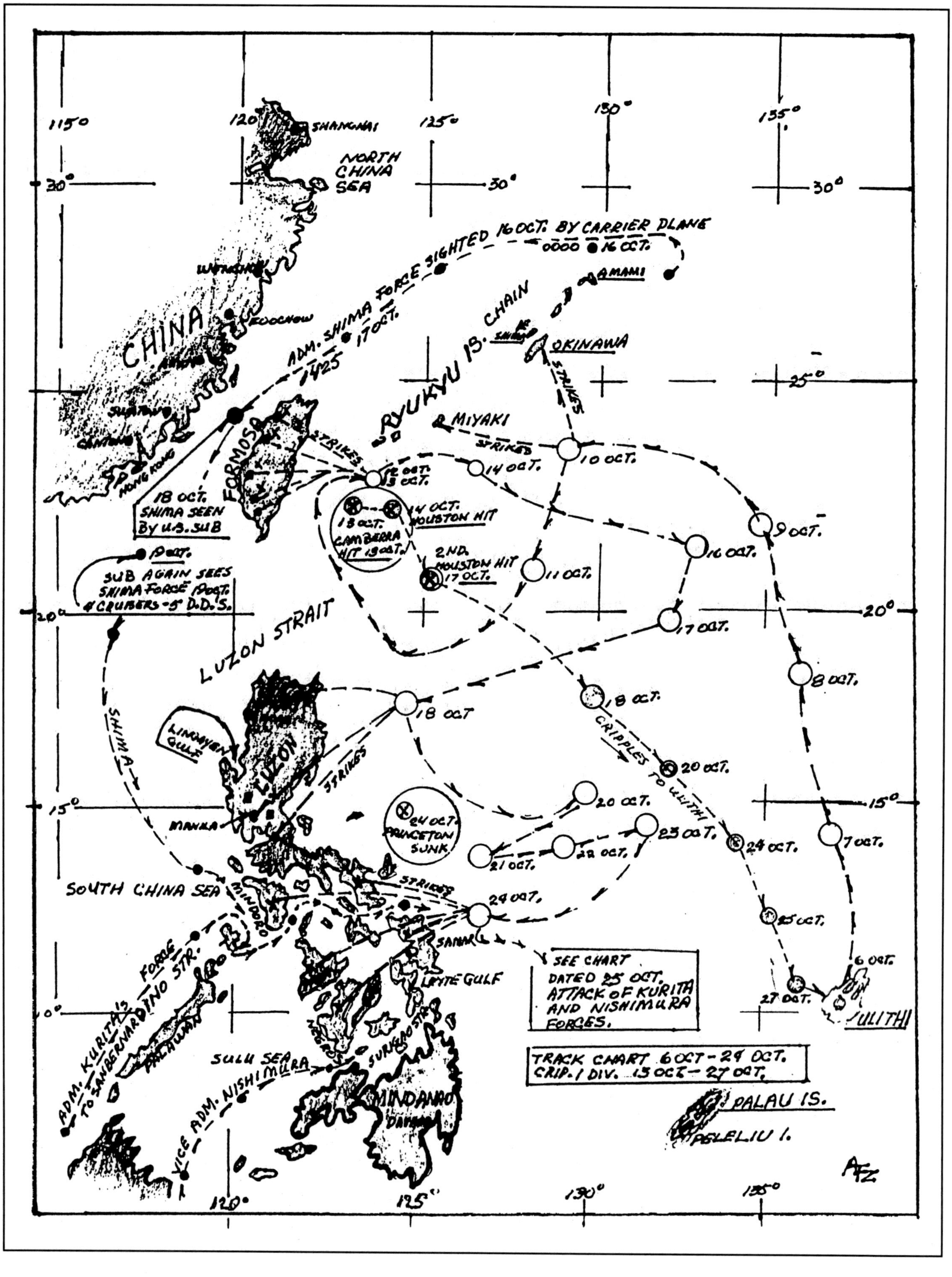

and big gun duel ensues as Oldendorf moves up his battleships and cruisers and a third line of destroyers. The night is lit up with shipboard explosions. At 0300 hours the battleship *Yamashiro* is torpedoed but remains in action. *Fuso* explodes and breaks into two pieces. A fourth line of U.S. destroyers now moves to the front and a group of nine launch 13 torpedoes at the *Yamashiro*. Four find their mark. The U.S. destroyer *Grant* is hit again and again by enemy gunfire and retired, in tow. *Grant* lost 34 men killed and 94 wounded. Nishimura now has left and floating are one damaged battleship (*Yamashiro* with Nishimura on board), one damaged cruiser (*Mogami*) and one destroyer (*Shigure*). It is now 0350 hours on the 25th of October 1944 and Oldendorf's battleships now are in range and fire on the remnants of Nishimura's force. Then Oldendorf's cruisers are in range and chime in. The U.S. destroyer *Grant* is hit by friendly fire in the darkness. Some explosions are now seen on *Yamashiro's* decks and before long she is fully ablaze and turns to make a run out of range. At 0400 hours she is hit again and at 0419 the *Yamashiro* rolls over and goes down taking the Admiral with her. Cruiser *Mogami* heads away from the area. Destroyer *Shigure* finds protection out of sight behind a projection of land and later follows the *Mogami*. Both escaped Oldendorf's immense firepower but both had much damage.

Admiral Shima, behind Nishimura's force, knows there is trouble ahead and is proceeding cautiously into the Mindanao Sea. He reaches the area of the P.T. boats and one P.T. sends a torpedo into his formation and it strikes the light cruiser *Abukuma*. One P.T. receives hits as Shima returns the fire. Several P.T. crewmen are wounded. Shima then runs across the two still floating pieces of the *Fuso*. It is 0430 hours and now aware of the devastation to Nishimura's force, Shima recalls his leading destroyers and begins a retreat. At this time the *Mogami*, on fire, comes down the Strait and in the confusion collides with Shima's heavy cruiser *Nachi* whose stern is badly damaged and her speed is reduced. *Mogami* joins Shima's force. Destroyer *Shigure* now comes along and she too joins Shima. It will not be daylight yet for over an hour. The crippled destroyer *Asagumo* also catches up with Shima making three ships of Nishimura's force which survive, although badly hit. The next morning U.S. surface ships finished off Nishimura's damaged destroyer, and Avengers from the escort carriers found and almost destroyed Nishimura's damaged heavy cruiser *Mogami* (the Japanese had to sink it). The following day Army Air Force bombers found and sank Admiral Shima's damaged cruiser *Abukuma*. We can now sum up the results of these actions against Admirals Nishimura and Shima. One of Nishimura's destroyers, the *Shigure* got away and arrived safely at Borneo on 27 October. She was the only ship to survive of Nishimura's force. Admiral Shima in his retreat rather than face Oldendorf's big guns, saved four destroyers and one heavy cruiser. Admiral Nishimura was killed.

Cape Engaño Action

On the morning of 24 October Admiral Ozawa's 17-ship northern force of four carriers, two battleships with flight decks, three light cruisers and ten destroyers came within range of T.F. 38 and launched 87 aircraft southward. They were driven off by CAP and the survivors landed on Luzon. The strike accomplished nothing. With this launch, the four carriers had remaining a total of 29 aircraft among them (of the original 116), 19 of them Zero fighters. Halsey launched scout planes and at 1640 hours on the 24th they found Ozawa's force less than 200 miles away. On the early evening of the 24th, Halsey ordered all his fast carrier task groups northward to hit the enemy carriers. This was precisely the Japanese plan: to lure the fast carrier fleet away from the Leyte area and give Kurita a chance to attack and disrupt the Leyte invasion. Ozawa's force was expendable and he expected to be attacked and to lose every one of his ships. He was nothing more than bait and Halsey bit. But, in fairness, Halsey did not know the weakened state of Ozawa's air strike capabilities, and he could not take any chances. He took the available three fast carrier groups with him.

Halsey's move leaves the San Bernardino Straits unprotected. Oldendorf was in position to attack Nishimura's southern force, which he did as we have seen, and he is not worried about Kurita's force as he believes Halsey's carriers have taken care of Kurita. At first, Oldendorf does not know that Halsey has pulled the fleet north to hit Ozawa, scheduled for the morning of the 25th. Also, Admiral Kinkaid, commanding the Seventh Fleet, is unaware of the fast carrier move northward, leaving the area unprotected. He believes that any force coming from the north to his southern position will be intercepted by the fast carriers and stopped.

Halsey had assumed that the previous actions against Kurita had weakened his force to the extent that Kinkaid could handle the situation should Kurita reappear. All very logical. *Intrepid* flag (Admiral Bogan) was about to complain that this action to drag the fleet north was not good planning. Bogan knew what was taking place at the straits and was about to contact Halsey and try to get him to leave *Intrepid* and 38.2 where they were to protect the straits and let only Sherman's and Davison's groups take on Ozawa. When Admiral Bogan talked to Halsey, one of Halsey's staff officers brushed him off with the reply that Halsey knew of the enemy ships and their heading. Admiral Lee also sends Halsey a message to try to get him to realize that Ozawa's force is a decoy. He too is given short shrift. Admiral Lee again contacts Halsey after scout planes verify the movement of Kurita's force eastward toward San Bernardino Straits. No reply from Halsey.

All of these messages to and from Halsey causes Halsey to send a message to Admiral Nimitz at Pearl and to General MacArthur explaining his plans on these late hours of the 24th. In his own defense of his actions of not hitting the two enemy forces to the south (Nishimura and Shima) and his reasoning in taking the force north, Halsey said, *"Searches of my carrier planes revealed the presence of the northern force of carriers on the afternoon of 24 October which completed the picture of all enemy forces. As it seemed childish of me to guard statically San Bernardino Strait, I concentrated Task Force 38 during the night and steamed north to attack the northern force at dawn (on the 25th). I believed that the center force (Kurita) had been so heavily damaged in the Sibuyan Sea that it could no longer be considered a serious menace to Seventh Fleet (Kinkaid)"*.

Admiral Mitscher, Task Force 38 Commander, who for days had been overstepped by Halsey, has not been consulted on the most recent decision to move the fast carrier fleet north. He is a little bitter about it all as he had had no part in these plans. At the time Mitscher retired for the night his chief of staff, Commodore Arleigh Burke, is in flag plot and talking to his assistants. Burke then says that he wished to check on the enemy forces to the south. He reads the report made by carrier *Independence* search planes at about one hour before Halsey pulled the fleet north. The report states that Kurita has swung around and is headed for San Bernardino Strait. Kurita is still very much afloat and intact and on the move. Burke feels that Admiral Mitscher should be told and he wakes the Admiral. He believes Admiral Lee's battleline (battleships and cruisers) which Halsey has taken north, should be dispatched to hit Kurita's force. Burke asks Mitscher to do just that, to inform Halsey to send battleline back to the south. Admiral Mitscher asks Burke if Admiral Halsey has that report; Negative answered Burke and Admiral Mitscher replied, *"If Halsey wants my advice, he'll ask for it."* Mitscher then went back to bed. Let's take a look at this force of Halsey's and his strike at Ozawa's northern force. We will return in a moment to the final battle with Kurita off Leyte which started out to be a near disaster for the U.S. forces there.

Halsey is steaming north on the night of 24 October with some 65 ships to hit Ozawa's group of 17. Admiral Mitscher is nominally in command of the three fast carrier groups, 38.2, 38.3 and 38.4. He orders that all planes be ready for a very early take off before dawn. The deck crews work through the night arming and fueling the planes and getting them spotted on the flight deck. The pilots and aircrews are fed at 0500 hours. The crew is fed at 0600 hours. The strike force takes off at 0540 hours and all are up by 0600 hours. Ahead of the strike planes are several search planes trying to spot Ozawa's force. His position was known at 0230 hours so the planes fan out from that position looking for Ozawa. At 0710 hours Ozawa is spotted and the strike force already in the air is given his position, about 200 miles east of the northern tip of Luzon. The two carrier forces are now 145 miles apart. The search planes fly over Ozawa's force and report back what the situation is and they direct the strike force right to Ozawa. The carrier planes arrive over Ozawa's force in one hour. This action would be known as the Battle of Cape Engaño.

It is a little past 0800 hours when the strike force of 180 planes begins its attack. The Helldivers go down first, followed by the fighters and then the Avengers move in releasing torpedoes. A destroyer is hit and blows up. Ozawa had asked for this strike and he got it; he was not taken by surprise, as he had watched the strike planes coming at him and he had his search planes up reporting back as the strike force closed the distance. The CAP over Ozawa's force was only a dozen planes which were swept aside. He had all hands at their guns and he believed he really could down many of the attacking planes. The carrier *Zuiho* leaves formation to launch her planes. Torpedo bombers from *Essex* and *Lexington* miss her but an *Intrepid* plane gets a hit on her. The carrier *Chitose* is hit several times and she has a ruptured side, takes on water, and goes down by 0930 hours.

The second strike force consisted of 16 torpedo planes, 6 bombers and 14 fighters. At 0951 hours 20 to 25 bogeys are headed to the U.S. task force. CAP heads out to meet them and they turn tail and run. With no CAP over Ozawa one plane is kept over the area all day to report the situation back to Admiral Mitscher. As the second strike hits, Ozawa's force seems confused; his ships are heading in all directions. There appears to be no control. These planes of *Lexington* and *Franklin* making this strike go after the carrier *Chiyoda* first. The carrier is hit and burning badly and out of control. She is taking on water and has begun a list. The battleship *Hyuga* tries to take the *Chiyoda* in tow but the strike planes interfere. Carrier commander Admiral Matsuda orders the light cruiser *Isuzu* and destroyer *Maki* to pull alongside *Chiyoda* and remove the crew. This can not be done due to all the action and confusion taking place. She remains afloat. A light cruiser is also hit by this strike.

The planes now go after the light cruiser *Tama* and the battleship *Hyuga,* with one destroyer close to her. The planes hit the cruiser. Strike number three is about to take off; it is just after 1200 hours. The tracker plane over the target area reports back that the two carriers *Zuikaku* and *Zuiho* are attempting to get away. They have one destroyer ahead of them and three others bring up the rear. They are at full speed. The battleship *Ise* is also with them. Also under way away from the area is the light cruiser *Tama*, damaged and leaking oil and making only ten knots. A few miles from her is the battleship *Hyuga* and one damaged destroyer. These last two are circling the damaged *Chiyoda.* Ten miles away is another light cruiser and still another cruiser also leaking oil. Behind these two is a destroyer dead in the water. Some 14 ships still afloat and most under way. All this is reported back to the carriers as strike three takes off. Strike three has orders to hit the main body of Ozawa's force and leave all the crippled and damaged ships for a later strike.

Strike three is the largest strike of the day. It is made up of planes from the first strike and others and numbers 200 planes. The tracker plane at the attack area is recalled and a fresh plane is sent to continue reporting back to the task force. The carriers *Zuikaku* and *Zuiho* are attacked in this wave. Three torpedoes hit *Zuikaku* and she burns. *Zuiho* is hit by a bomb and is also set on fire. *Zuiho's* fires are being brought under control and she turns to run at a high rate of speed. *Zuikaku* is beginning to settle in the water. As the planes are about to return to their carriers the order is given to hit the fleeing *Zuiho*. Some 40 planes go after her at 1300 hours and again at 1330 hours and she is greatly damaged. The *Zuikaku*, last of the Pearl Harbor strike carriers, now rolls over and does down at 1415 hours.

Strike number four, only 35 planes, is on the way after a take off at 1315 hours. They arrive over the area at 1445 hours. Ozawa's fleet is now well broken, and scattered over some 30 miles of ocean. The battleship *Ise* is hit several times. Some two dozen planes go after the crippled *Zuiho* and she goes down at 1530 hours. The fifth attack — 96 planes — took off at 1610 hours This is made up from each of the five carriers, and they go after the battleship *Ise*. This ship had many AA batteries and they were fairly accurate. The *Ise* fought off the attackers — she was not about to be taken. She maneuvered every time a plane made a run on her and she avoided being hit. At this late hour of the day Ozawa's ships are scattered, some trailing oil, others at slow speed. A sixth strike of 36 planes takes off at 1710 hours to hit the ships a final time. About 12 of Ozawa's ships are still afloat. More hits were made but no sinkings. All in all, the work of the carrier planes for the day was generally of low quality and disappointing for all the effort expended. Five ships of the 17 ship force were sunk, including three of four carriers. Halsey then sent a cruiser and destroyer force north at high speed to engage the remnants of Ozawa's retreating fleet in a surface battle. The caught up with him and sent the last carrier, the badly damaged *Chiyoda*, and a destroyer to the bottom. The cruiser-destroyer force then broke off the engagement. U.S. submarines later sank a damaged light cruiser. The rest of Ozawa's force — his two battleships, two light cruisers and six destroyers — escaped.

Now we must return to events leading up to the final actions off Leyte. At about 0830 hours on the 25th as strike number two is returning from its strike on Ozawa's force, the radio breaks silence and Admiral Kinkaid is sending a plain language message. He is informing Halsey that his fleet is under attack off Samar and is pleading for assistance. (Kurita had steamed through the San Bernardino Strait, turned south and ran into the northern most group of the Taffys). Admiral Halsey does not realize that he left the area unprotected when he pulled the fleet north. On the *New Jersey*, Halsey's flag ship, he listens to Kinkaid's pleas. Halsey ponders the situation for a while then orders Admiral McCain (T.G. 38.1) to head to Samar at the best possible speed to assist the Taffys. McCain was refueling 400 miles east of Leyte at this time. The time of this order is just before 0900 hours, already too late. Halsey does not dispatch the fire power group (to be called Task Force 34) until later and later was too late as Kurita would escape.

Back in Hawaii Admiral Nimitz is listening to this and other messages flying between Kinkaid and Halsey and now is a little perturbed about the way things are being handled. It is now 1000 hours as Halsey receives another message. It is Admiral Nimitz calling. His tone of voice indicates some uneasiness as he asks, "*Where is Task Force 34*?" Halsey replies that he took them with him to use to chase down the ships that might get away and to sink those that are disabled. There was in fact no Task Force 34 until Halsey decided to combine Admiral Lee's big guns (battleships and cruisers) sailing with him with carrier group 38.2 (*Intrepid, Cabot* and *Independence*), also in this force pursuing Ozawa. A few minutes before 1100 hours on the 25th of October Halsey ordered Lee's "battleline" — six fast battleships — and T.G. 38.2 to reverse course and steam full speed to the Taffys at Leyte. They are to assist Admiral Kinkaid's Seventh Fleet in any way they can. The order is executed at 1115 hours. From where the combined fleet now is located it would be early the next morning before they could arrive at San Bernardino Strait. Much, much too late to assist.

Halsey was somewhat angered when he had to pull the battleline away from the excellent targets they had of Ozawa's crippled force. Halsey later will write of these late decisions and how all was effected by them. He will write that because Nimitz was so worried over the fate of Kinkaid's Seventh Fleet that this influenced his decision to send back 38.2. Admiral Lee with battleline must slow down to refuel the smaller ships in his group as he heads south. Having been making 20 knots he now is down to 12 knots and refueling.

It is 1345 hours. Admiral Halsey now forms Task Group 34.5 with Rear Admiral Badger in command. It is 1630 hours. At 1700 hours 34.5 heads south toward San Bernardino Strait at 28 knots. This force has the battleships *Iowa, New Jersey,* three light cruisers and eight destroyers. Admiral Badger and T.G. 34.5 were to arrive off the strait at 0100 hours, 26 October. Admiral Bogan's 38.2 (*Intrepid*) would be east of the strait to provide air cover for them. This force is to sweep the areas of the strait then sweep the coast area of Samar south to Leyte Gulf. The orders are to destroy anything and everything in the areas. All would be for naught. Kurita, after attacking Taffy Three, would by this time have re-entered San Bernardino Strait and escaped westward. Now we should look at Kurita's sortie off Samar, and for the U.S. it is not a pretty story.

Samar Action

This final engagement, known as the Battle off Samar, began on the morning of 25 October, as Halsey was attacking Ozawa's carrier force several hundred miles to the north. Kinkaid's Taffy Three was off Samar protecting the northern approaches to Leyte Gulf. The other two Taffy groups were to the south, Taffy Two off Leyte Gulf and Taffy Three to the east of Mindanao. The three groups totalled 14 escort carriers carrying about 38 to 40 planes each. (Two of the escort carriers of Taffy One were en route to Morotai to pick up replacement aircraft). 21 destroyers screened these groups. The mission of the Taffys was to provide air cover over the Leyte landings and maintain a CAP over the area and over the ships as well as ASP (anti-submarine patrols). They also had the responsibility of attacking enemy convoys and destroying enemy installations such as ammo dumps and oil installations around the invaded areas, in this case, all of Leyte Island. An hour or so before sun up on 25 October the orders are to launch strikes to hit different areas. The most southern group, Taffy One with four carriers, has orders to locate and destroy the enemy force moving westward from Surigao Strait (the remnants of Nishimura's force). The northern most group, Taffy Three with six carriers, was to be the cover for the ships and she launched the combat air patrol (CAP) group. She did send about six planes on the strike and had other planes on ASP. Directly east of Leyte Gulf, Taffy Two also with six carriers sent off about ten planes to join Taffy One planes in the search and strike mission. The three groups were 30 to 50 miles apart.

Although Admiral Kurita's fleet had been badly mauled at Tablas Strait on the 24th, and he had turned away from San Bernardino Strait, his retreat was only temporary. Third Fleet commander Halsey thought his carrier planes had put Kurita out of action more or less permanently. His information was faulty. Kurita's retreat was only temporary, for that same night he headed back toward San Bernardino Strait after Halsey had pulled all his fast carriers north after Ozawa. Kurita was lucky. When his force exited the Strait in the early morning darkness of 25 October there wasn't an American ship in sight and he immediately turned east and then south along the coast of Samar, headed for Leyte Gulf. His force was still formidable: four battleships, six heavy and two light cruisers, and thirteen destroyers. At this time Kurita had no knowledge of the position of any American force but he did know of the fate of Nishimura's force in the Mindanao Sea — or at least that he had come out second best.

Taffy Three was commanded by Admiral Clifton A.F. Sprague (no relation to Admiral Thomas L. Sprague, *Intrepid's* first captain, who was in overall command of the all three Taffy groups). Cliff Sprague's Taffy Three was undergoing routine flight maneuvers early on the morning of the 25th when his lookouts saw flak bursts to the northwest. No one was supposed to be north of Taffy Three. Not within sight, anyway. An Avenger on patrol had spotted Kurita's force steaming south along the Samar coast, identified it as the enemy, and attacked. In a few minutes the tell-tale pagoda superstructure of Japanese battleships and cruisers poked over the horizon. The two forces sighted each other at the same time, about 0645 hours, 25 October.

Kurita, as he approaches Sprague's small flat tops, thinks perhaps he has stumbled upon the Task Force 38 fast carrier force. Another thought crosses his mind and that is: Is this Ozawa's carrier force this far south? Kurita is making about 20 knots and closing the distance on the small carriers. The attack by the Avenger made it obvious it was not Ozawa. Kurita formed up for battle. The weather is now overcast and visibility is decreasing. Kurita has spread out his force and is approaching in four columns There is nothing weak about this force as Admiral Halsey had so readily believed. In this force is the giant *Yamato,* a so-called super battleship with 18 inch guns. It is Kurita's flag ship. His battleship *Nagato* has 16 inch guns and the battleships *Kongo* and *Haruna* have 14 inch guns. His two light cruisers are the *Nashiro* and the *Yahagi.* This force had moved down the Samar coast to this point undetected and it worried Kurita. This opportunity seemed too good to be true; all his fire power against smaller and slower carriers that had only one 5 inch gun each — and of course, their planes. But against these, he had almost impenetrable anti-aircraft defenses.

Kurita orders his big guns to begin their shelling of the escort carriers — there are six of them with their destroyer escorts. It is about 0700 hours. Admiral Sprague turns his whole group east, lays down a smoke screen, and launches his planes. Cliff Sprague is in trouble; Kurita can make close to 30 knots with his big ships and an escort carrier is lucky to make 17 with a full head of steam. There is no way Sprague can run away from them. Sprague sends out an urgent call for help. Kurita's battleships are leading the force. They come on in groups of two as Kurita gave the attack order. The first two ships to approach Sprague are the *Yamato* and the *Nagato.* Behind these battleships are six heavy cruisers and behind them the destroyer screen. This is a strange formation; the destroyers should have been in the lead, not in the rear. The heavy cruisers are spread out in three groups of two each. In the center of the destroyers are the two light cruisers.

Things are looking dark for Clifton Sprague and his small flat tops. Admiral Tom Sprague is some 130 miles away with Taffy One. Admiral Stump is closer to Admiral Clifton Sprague's Taffy Three with his Taffy Two. Stump hears Cliff Sprague's call for assistance and acts quickly to get permission from Admiral Kinkaid to launch all aircraft and go to the aid of Taffy Three. By now Sprague has enemy ships on his quarter and coming up at the rear. Taffy Three has only one way to go and that is to head for Leyte Gulf as his planes, now in the air, tangle with Kurita's heavy fire power ships which are now all bunched in a small area. Admiral Stump then comes on the TBS (short range Talk Between Ships) and assures Cliff Sprague he is not far away and to hang on. Sprague had already changed course, got his planes in the air, made a lot of smoke and like the old days out west when the wagons formed a circle, so too, Clifton Sprague orders the carriers into a circle.

Kurita now opened up his big guns on the escort carriers. C. Sprague's flag ship, *Fanshaw Bay* and the *White Plains* are in the line of the first salvos from the battleships. The *White Plains* takes several near misses from 14 inch shells and she momentarily looses steering control. An electric generator circuit had opened but the problem was quickly found and corrected. Taffy Three had its destroyers deployed around the circle of carriers and orders all of them to launch torpedoes at the oncoming enemy fleet. At this time a fast moving rain squall moves in and the small carriers break formation and head into it. By now the carrier planes have begun to attack Kurita's ships and at the same time Cliff Sprague alters course and heads in the direction from which soon should appear Taffy Two's planes and perhaps the planes from Taffy One.

It is now 0735 hours, all this action and movement having started about 0645. Kurita does not alter his course and he has increased the distance between him and Sprague's carriers. The fire from Kurita's ships now fell short of the small carrier force. Clifton Sprague had bought precious minutes with his quick moves. Kurita's force is now broken up with distance between his ships and Sprague's and now destroyers are moving in between him and Sprague and launching torpedoes. Kurita is now worried about the outcome since he gave the order to attack without putting his ships in an attack position, and also his destroyers brought up the rear instead of being up front in point position ready to launch torpedoes. The rain squall also bothered him as it allowed Clifton Sprague to enter it and then reverse his course opening the distance between the two fleets.

Destroyer *Johnson* is first to maneuver into torpedo firing position, her deck guns blazing away even before her torpedoes are launched. The *Johnson's* target is believed to have been the heavy cruiser *Kumano* which was the lead ship in a column of several. The *Johnson* is now the target of several of the big gun ships. The destroyers *Heerman* and *Hoel* are now also in a launch torpedo position. These three destroyers are of the new 21,000 ton *Fletcher* class and the *Johnson* is just one year old as this action begins. The *Johnson* is closest to her target and lets go all ten of her torpedoes and instantly turns, making smoke and pulls away. She fired her torpedoes at the cruiser *Kumano* which lay 10,000 yards away from her. Two explosions are heard but not seen through the smoke screen. As she comes clear of smoke, she reports the *Kumano* burning bad aft of her bridge. The heavy cruiser *Suzuya's* speed has been reduced to 20 knots or less by a bomb hit from one of the carrier planes.

It is now a little past 0730 hours and the *Johnson* pulls away after launching her torpedoes. Suddenly three 14 inch battleship shells hit her and she takes several more from a cruiser. These hits knock out the engine room and she looses all power to her steering and power to her three aft 5 inch guns. Her gyro compass is rendered useless. Another shell snaps off her mast and the radar rig fell on the bridge killing three of her officers. Her deck is opened and men below are killed. With the after engine room out, her forward engine gives the *Johnson* speed up to 17 knots. Power is restored to two of the three aft 5 inch mounts. Steering is now manually operated.

The squall which saved the carrier in the initial attack now rolls over the *Johnson*. But as she comes out of the squall she is hit again; many more shells cripple her and she goes down. The destroyer *Hoel* was also lost but before she went down she got into position to torpedo the battleship *Kongo*. This action occurred some seven miles away from the *Johnson's* last position. The *Kongo* opens fire on the *Hoel* as the destroyer lets go four torpedoes at the battleship. *Kongo's* shells hit the *Hoel* and her after fire room and her turbine is knocked out, her steering is out as her rudder is hard over, and control to three of her guns is also gone as is one of her engines. Unable to steer, the destroyer is headed straight to her target, the battleship *Kongo*. The destroyer then blew up and went down. It is reported that *Hoel* took some 40 hits of all caliber including 5 inch, 8 inch and 16 inch shells before she sank. (The *Kongo* survived this action but she would be sunk in a matter of days — 1 November at Foochow — by a submarine).

It is now 0800 hours and Admiral Sprague orders a second torpedo run on Kurita's force. The destroyer *Heerman* has yet to fire her torpedoes and now she prepares to do just that. By now the rains come and with all the smoke of the big guns and the screens visibility is poor. *Heerman* moves up to attack the cruiser *Haguro* but her torpedoes miss and the *Haguro* opens up on the destroyer. To the port of *Heerman* is the battleship *Kongo* . To the *Kongo's* stern is the battleship *Haruna*. Further back are the battleships *Yamato* and *Nagato*. Commander Hathaway of the *Heerman* is determined to face all these ships and, altering his course slightly, he put himself into position to hit the nearest ship. Hathaway fired three torpedoes at the battleship *Haruna*. The fast moving *Heerman* then turned and headed to the carrier formation. One torpedo hit the *Haruna* but damage was slight. The battleship *Yamato* was attacked by destroyers and she too evaded a half dozen torpedoes. So far, the U.S. had lost two destroyers and had inflicted only minor damage on Kurita's ships.

When the order was given to launch the second torpedo attack several destroyer escorts (D.E.) along with the *Johnson, Hoel* and *Heerman* also made runs on the Japanese ships. They were the *Roberts, J.C. Butler, Dennis,* and *Raymond*. Their action occurred about the time the *Johnson* was hit. The *Roberts* found herself in range to launch torpedoes at the huge *Yamato* but that battleship was not about to be hit by a little D.E. The *Yamato* opened fire on the *Roberts* just as the *Hoel* and *Heerman* launch their torpedoes. *Yamato* evades every torpedo aimed at her. The *J.C. Raymond* now comes up to take a crack at the big boys and launches her torpedoes just as the *Johnson* is hit. The *Raymond* positions herself to fire at the cruiser Haguro. The cruiser opens fire on the *Raymond* as the D.E. launches three torpedoes. At this moment an Avenger drops a torpedo at the *Haguro*. The cruiser altered her course and all the torpedoes miss her. The cruiser's guns are now trained on the attacking destroyers and the D.E.s and her range is such that she can't help but hit one of her attackers. Time now is about 0750 hours as the D.E. *Dennis* moves in on the *Haguro* and the cruiser is now firing on both the *Dennis* and the *Raymond*. The *Dennis* launches her torpedoes not at the *Haguro* but at the cruiser *Tone*. Everything missed. Bombers now single out the *Haguro* and she is hit and her No. 2 turret is out of commission. All these attacking D.E.s make smoke and return to the carrier formation. They are unharmed but so is everything they attacked.

The escort carriers meanwhile continue to move in and out of the rain squalls and have managed to avoid being hit by the fire of Kurita's big guns. But Kurita is now closing in on the carriers and Admiral Sprague orders a third torpedo attack. By this time some of the destroyers are out of torpedoes. The D.E.s *Butler, Dennis* and *Raymond* respond to the call and make runs on the cruisers *Tone* and *Chikuma*. The *Chikuma* lays down gun fire on the *Butler* who puts out smoke and pulls back. The *Dennis* is also under attack and she is hit twice below the waterline and joins *Butler* behind the smoke. The *Raymond* is also on line and firing at Kurita's ships. The big guns are now close enough to turn on the baby flat tops. The *St. Lô* is the nearest carrier so Admiral Sprague orders *Butler* to position herself just forward of the *St. Lô* and make smoke. It is now just after 0900 hours and the *Roberts* is hit below the waterline. She then takes several more hits. At 1000 hours she goes down, her side ripped open by big shells, probably from the *Kongo*. Three officers and 86 men are killed; this is half her company.

The cruisers *Tone* and *Chikuma* are now under attack by planes from Admiral Stump's Taffy Two. The *Chikuma* turns and withdraws. The *Tone* is also being attacked by the destroyer *Heerman*. Then the *Tone* turns and follows the *Chikuma* away from the action. Later the reports will state that many enemy torpedoes were blown up in the water by attacking Taffy Two planes, before they found their American ship targets. The gallant *Johnson* in her last moments put up a good fight. As she lay dead in the water she was pounded continuously until she went down. Of her crew of some 327, only 141 survived. In the final count, almost all torpedoes fired at the enemy missed their mark except for the one the *Johnson* laid on the *Kumano*. Kurita will report later that the destroyer actions broke up his force by their continued fast movements in around his ships as they let go some 30 torpedoes. He also reported that the smoke screens put out by the American ships bothered him almost as much as the torpedoes coming at his ships.

It is time now to try to report on what the baby flat tops were doing during the action packed hours of 0700 to 1000 hours on this 25 October. The escort carriers are moving three abreast at about 17 knots as they came out of that early morning rain squall. Bearing down on them are four heavy cruisers, the *Chikuma, Tone, Haguro* and the *Chokai* approaching from the port side. It was at this time that the small carriers' cover force of destroyers and destroyer escorts tangled with the big gun ships and in the process let go some 30 torpedoes. With four cruisers coming on from port and additional cruisers and battleships to his rear, and a group of Kurita's destroyers coming on from the starboard, Admiral C. Sprague is essentially surrounded. The wind is to his rear and he is launching planes as he moves out of the rain squall. He orders the planes to attack the four cruisers first. He orders the carriers to open fire upon the cruisers also. The *St. Lô* opens fire even before the order is given. The carriers commence evasive action and they keep changing course as the big shells of the cruisers drop all around them. It is a few minutes before 0800 hours and the *Kalinin Bay* is hit. She responds with her single 5 inch gun. She hits the second turret of one of the heavy cruisers, putting it out of action as at the same moment she takes an 8 inch salvo from the cruiser.

The small carriers make a lot of smoke and so do the destroyers. The *Kalinin Bay* and *Gambier Bay* bring up the rear and the wind is such that the smoke offers them little cover as the smoke in front of them blows away from them leaving the two fully exposed to the cruisers at their rear. The Japanese gunners were not too good as many shells landed short while other salvos overshot their targets. Their range was inaccurate as was their gun deflection. They fired high, wide, in front of the target or to its rear, everywhere but on the target. They are aiming at the smoke screens and the zigzagging small carriers added to their inaccuracy. As the big guns of Kurita's ships continue to send salvo after salvo at the small carriers the flag ship of Admiral C. Sprague takes a hit and is hit four times with two very near misses. She has three men killed and 20 injured. The *White Plains* takes a minor hit also and has only slight damage. *Kitkum Bay* is not hit directly but has some very near misses which give her some wounded men, hit by hot steel. *Kalinin Bay*, first hit, now has taken a dozen hits and is taking on water and has no steering. Her crew is below trying to stem the flow of water into her while others are aft trying to steer by hand power below the decks.

Most of Taffy Three's aircraft are launched by 0730 hours, and the carriers of Taffy Two, some 65 miles to the south is also launching their planes. Taffy One with four carriers is another 65 or so miles further south. The first Japanese ship hit was the heavy cruiser *Suzuya* which loses power and slows. The planes made so many passes at the enemy fleet that they ran out of ammunition and bombs and torpedoes. They then made dry runs to divert the enemy ships' courses and interrupt their fire on the small carriers. The fighters also expended all their machine gun ammo trying to strafe and knock out the guns of the enemy ships. Some planes made six, seven, eight, up to a dozen dry runs to divert the enemy ships' fire.

At about 0730 hours the escort carrier *Gambier Bay* is under attack by Kurita's big guns. Not until about 0800 hours did she take her first hit, a shell exploding on her flight deck, aft, holing the deck and starting fires on her hanger deck below. The carrier soon takes several more hits, as the enemy's big guns now concentrate on the smoking ship. She has a hole in her side below the waterline and the engine room is flooded. She loses speed and power and falls away from her sister carriers, now fully exposed to the enemy's guns. *Gambier Bay* is afire now and her 20 degree list is increasing. The heavy cruiser *Chikuma* opens up on the stricken carrier as other cruisers also begin to use her for target practice. At 0845 hours the *Gambier Bay* is dead in the water and about to go down. At 0850 the order to abandon ship is given. The little carrier went down at 0910 hours. Of her 854 crew about 750 escaped the burning ship and watch the *Chikuma* pound their carrier; some of this fire power sprays shell fragments over the men in the water and some are killed. These men will be in the water for over two days before they are picked up and several are pulled under by sharks.

The escort carrier *St. Lô* is also under attack as the *Gambier Bay* is being hit. She has the enemy on both sides of her from the beam aft and others to her rear. The *St. Lô* begins to make frequent course changes, zigzagging, never staying still. The shells drop on either side of her and beyond her. The 5 inch 38 gun of the *St. Lô* is aft on the fantail and this gun begins to fire on her big targets off her quarter and to her rear. The *St. Lô* gun makes several hits on her attackers, six hits on a heavy cruiser alone. One of these hits puts a forward gun of the cruiser out of business. The cruiser was the *Suzuya,* and she is hit again and again by the *St. Lo's* 5/38. In less than 30 minutes the *Suzuya* was stopped, her engines out of commission. The accuracy of the carrier's gunners was unbelievable. At the same time the carrier planes are circling to make a run on the *Suzuya* and other targets. Four torpedo bombers dive on the heavy cruiser and in no more than a half minute score five hits amidships, three at the bow and one hit and two near misses at the stern. Another plane then makes a second direct hit on her stern which sent the cruiser into a sharp right turn. The *Suzuya* coasted some 500 yards further and then blew up and went down. It was 0930 hours. At this time all six carriers of Taffy Three had been hit and one sunk.

The ships at this time are in view of Samar about 30 miles distance. Admiral Sprague of Taffy Three had this to say when the action ceased, "At 0925 hours my mind was occupied with dodging torpedoes when near the bridge I heard one of the signal men yell, 'Goddamit, boys, they're getting away!' I could not believe my eyes, but it looked as if the whole Japanese fleet was indeed retiring. However, it took a whole series of reports from circling planes to convince me. And still I could not get the fact to soak into my battle numbed brain. At best I had expected to be swimming by now." In other writings and reports of these actions I came to the conclusion that Admiral Halsey also heard Admiral Sprague's call for help just prior to this action, and knowing that he never established Task Force 34 which would have protected this Leyte area, sent Kinkaid a message which in short said, "I am on the way — be there in two hours." Did Kurita also hear this message and is that why he pulled away unexpectedly? We will follow Kurita's withdrawal a little later. First we will detail more of this action of the 25th of October by looking at the effects of the enemy's first introduction of a new weapon — the kamikazes.

Taffy Three, although badly mauled by Kurita's big guns, now had a brief breather as Kurita withdrew. All six of Admiral Sprague's escort carriers had been hit and he now had only five left with the sinking of the *Gambier Bay.* Further south, about 130 miles, Taffy One was also having its troubles. Admiral Thomas Sprague was in overall command of all the Taffys and he was with Taffy One with its four carriers. Screening these carriers were three destroyers and four destroyer escorts. Although most of their planes were on a strike against the remains of Nishimura's force in the Mindanao Sea on this morning, T. Sprague sent off what planes he could to the north when he got word of Kurita's attack on C. Sprague's Taffy Three. Taffy One's *Santee* had just launched her fighters and torpedo bombers when an undetected Japanese fighter dived out of a cloud straight at the carrier. At 0745 the bomb-laden Zero plunged into the flight deck of the *Suwanee,* penetrated through to the hanger deck and exploded. When the flames were under control sixteen of her crew were counted dead and 27 injured. Enter the suicide pilot. The kamikaze corps had been organized only five days earlier and this was their first strike.

Seconds later a second suicider, this one hit by anti-aircraft fire from the *Suwanee*, dived on the *Sangamon.* He missed by about 500 feet, crashing into the sea. A third kamikaze, also smoking from a flak hit, just missed the *Petrof Bay.* A fourth Zero now aimed at the *Suwanee* and this one broke through the wall of flak and crashed onto her flight deck, sending her bomb into the hanger area. The explosion killed 70 and injured 80. To add insult to injury the *Santee* now took a torpedo from a Japanese submarine. There were no casualties from this hit and the little carrier kept formation. Finally, these brief few minutes of hell were over for Taffy One.

Further north, hell would begin again for Taffy Three as Kurita withdrew his forces. Five Zero suiciders found C. Sprague's mauled carriers off Samar and went in for the kill — of the carriers and themselves. Zero 1 dived on the badly damaged *Kitkun Bay,* bounced off her port catwalk and splashed into the sea. The exploding bomb caused more damage. The little carrier had already taken 15 shell hits. Zeros 2 and 3 dived together at the *Fanshaw Bay* and both were shot down in flames. Zeros 4 and 5 dived for the *White Plains*, were hit by flak fire sending one into the sea, its exploding bomb injuring 11 men on the *White Plains*, but altering the course of the other so it went screaming into the hapless *St. Lô.* The time was now about 1050 hours. A series of explosions now began tearing apart the *St. Lô.* She lost water pressure and her fires now raged out of control. More explosions as the carrier shivered and began to fly apart. The order was given to abandon ship and at 1130 hours she went under with 100 of her crew. The first sinking by a suicide plane. It would not be the last.

Then, shortly after 1100 hours, 15 Judys dived on C. Sprague's remaining four carriers. Three got through the carriers' flak screen and a few defending American fighters, to further damage the *Kalinin Bay* and the *Kitkun Bay.* Both carriers continued operations. The final blow was struck the next day, the 26th when, at about noon, many Zeros and Judys made one more run at T. Sprague's Taffy One. U.S. fighters shot down a dozen of them but one got through to crash into the flight deck of the *Suwanee*, already damaged on the preceding day. The resulting explosions and fires destroyed ten of the carrier's aircraft and killed more than 140 men. All told, the *Suwanee* sustained in these two days about 400 casualties — over a third of her crew. This action finally spelled the end of the ordeal for the Taffys. There was now some mopping up to do of Kurita's force, heading west from San Bernardino Strait.

At about 1300 hours on the 25th, several hours after Kurita broke off his attack on Taffy Three, about 100 planes from Admiral McCain's Task Group 38.1, which was retiring to Ulithi for rest but ordered back when Kurita struck, found and attacked Kurita's big gun ships. A second strike was sent out almost immediately, of 47 planes. These strikes made one bomb hit on Kurita's 22-ship force, and that bomb was a dud. They inflicted no damage and lost 14 planes in the attempt. These aircrews in fact were already exhausted from nearly ten months of continuous combat with the fleet. They just were not sharp any more. Late that night Kurita slowly wound his way through San Bernardino Strait and into the Sibuyan Sea. A night flying plane from *Independence, Intrepid's* Group 38.2, which had come south with Halsey's fast battleships from the Ozawa strikes, spotted Kurita on his radar and radioed back the sighting. Kurita had already navigated the strait when Halsey and his battleships showed up to find a destroyer straggler that had lagged far behind Kurita's main force. Halsey's ships sent the enemy destroyer to the bottom in quick order.

At 0500 on the 26th Task Groups 38.1 (McCain) and 38.2 (Bogan) rendezvoused off Samar and launched strikes to find Kurita. Over 250 planes converged on Kurita's force and succeeded only in sinking a light cruiser one destroyer, and forcing another destroyer to beach herself. The Battle of Leyte Gulf was over and the escort carrier pilots of the Taffys were the ones who had made the difference. It was an American naval victory to be sure, but it could have been much more disastrous for the enemy. Admiral Kurita had saved four of his battleships and some cruisers and destroyers; Admiral Shima still had two cruisers; Admiral Ozawa had two battleships with short flight decks, two cruisers and a few destroyers. There were few operational enemy aircraft carriers left and those still floating had few planes and fewer pilots to carry. These ships comprised the remains of the Japanese Imperial Navy. Ozawa had lost almost all of his original 116 carrier planes. U.S. fast carriers (Task Force 38) had lost a total of 40 aircraft in the three days of fighting, and the escort carriers (the Taffys) about 100 including those that went down with the two lost escort carriers. The American victory was clear but not clean. A comparison of ships sunk and damaged in these complex battles is of interest since it shows much more Japanese tonnage sunk than American tonnage. Damaged ships would be repaired to fight again.

Account of Damages of the Battle of Leyte Gulf

United States	**Japanese**
Ships sunk:	Ships sunk:
CVL-23 *Princeton*	CV *Zuikaku*
CVE-73 *Gambier Bay*	CVL *Chitose*
CVE-63 *St. Lô*	CVL *Chiyoda*
	CVL *Zuiho*
2 Destroyers:	
Johnson	3 Battleships
Hoel	6 Heavy cruisers
1 Destroyer Escort:	3 Light cruisers
Samuel B. Roberts	8 Destroyers
1 P.T. boat #493	
Major damage:	Major damage:
CVE-70 *Fanshaw Bay*	2 Heavy cruisers
CVE-68 *Kalinin Bay*	1 Light cruiser
CVE-71 *Kitkun Bay*	
CVE-29 *Santee*	
CVE-27 *Suwanee*	
CVE-66 *White Plains*	
1 Light cruiser	
3 Destroyers	
1 Destroyer Escort	

By night fall of the 26th it was no longer possible to pursue the remnants of Kurita's fleet and the strikes were called off. All groups of Task Force 38 now make plans for taking on fuel. Task Groups 38.3 and 38.4 are first to refuel. Other tankers are to meet 38.1 and 38.2 at another location. Early in the day Kinkaid asks Halsey to provide CAP over Leyte Gulf. Halsey replies he cannot since he is hours away north and refueling the ships. He can provide two groups on 27 October and they

Battle of Leyte Gulf

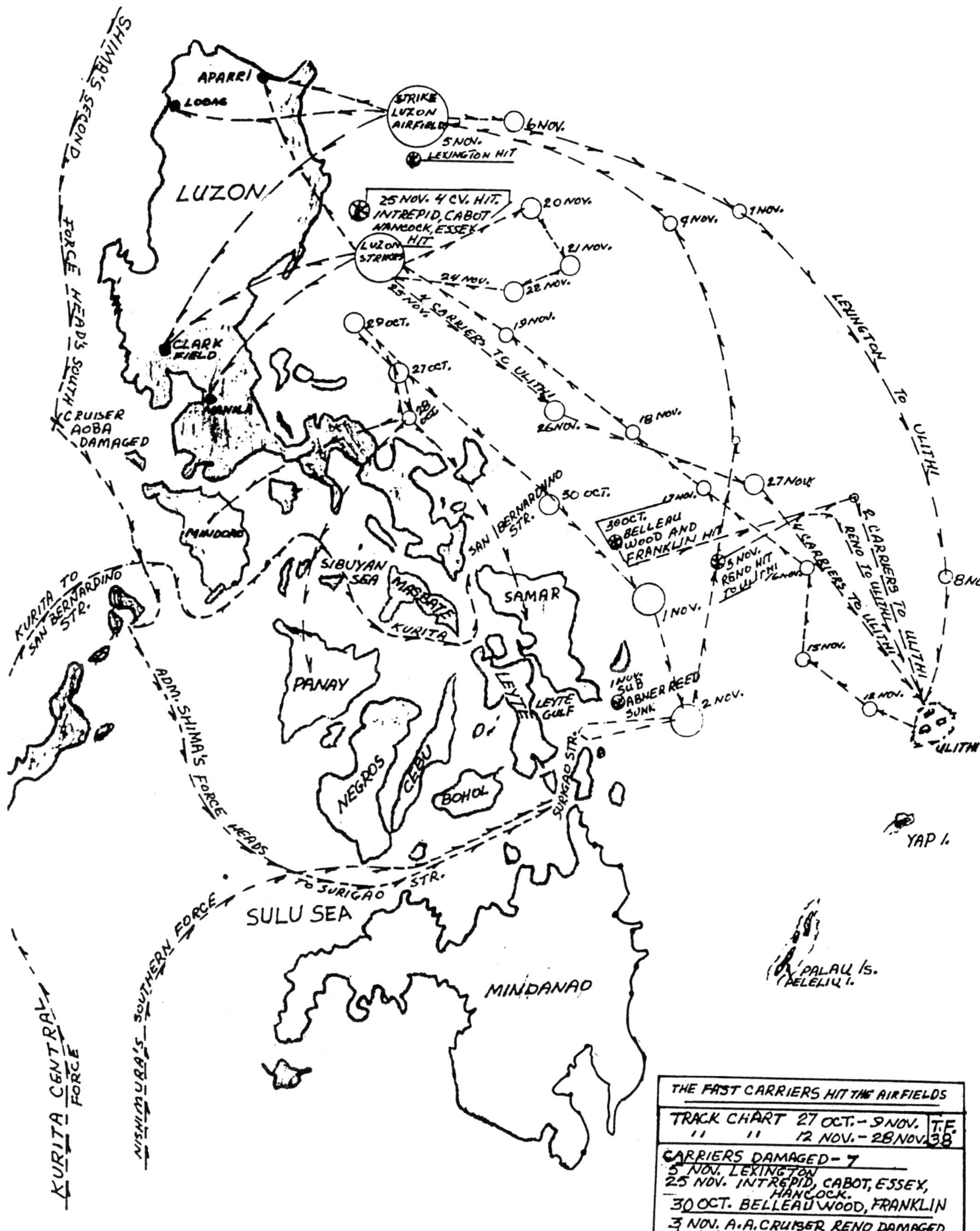

INTREPID IS THERE

would be 38.3 and 38.4 and they would assist Kinkaid with strikes or patrols which ever he required. He contacts Admiral Mischer and relays those instructions to him as he is now in command of 38.3 and 38.4. Along with refueling all ships take on ammo, bombs and torpedoes as those items are also almost depleted. *Intrepid* refueled on the 27th taking on "9,886 bbls. of fuel oil at 85°F and 81,600 gallons of AvGas." It took four hours.

Admiral Sherman's group 38.3 makes sweeps over Leyte Gulf and the Visayas and areas of Luzon being directed by Admiral Kinkaid. Carrier *Essex* of 38.3 locates a small troop convoy and sinks two of its protecting destroyers. They were the *Fujimami* and *Shiranuhi*. This was at the Calamian Islands. Admiral Bogan in *Intrepid* (38.2) after refueling joins Admiral Davison's 38.4 and together they give support to the forces at Leyte by striking the enemy air fields there. Admiral McCain's 38.1 is now able to get his long delayed service period for his group which had been called back to the Leyte battles. Sherman's group 38.3 is also dispatched with McCain and both head for Ulithi.

Remaining off Leyte and Samar were Groups 38.2 (*Intrepid, Cabot, Hancock* and *Independence*) and 38.4 (*Franklin, Belleau Wood, Enterprise* and *San Jacinto*). A raid is made on shipping at Cebu. The group is attacked by some 45 enemy planes and CAP shoots down 13 of them. 38.4 loses four aircraft and three pilots in this action. The weather turns bad in this area. A few planes are lost due to weather in trying to land. 38.4 now comes under attack by enemy submarines. Destroyers *Helm* and *Gridley* go after one sub who is only some 600 yards away. The ships pick up speed and move away from the area. Depth charges are dropped over the sub and hit it as all kinds of debris comes to the surface. More depth charges are dropped so as to be certain they got the sub and this time parts of the sub itself are shot to the surface, and a very powerful explosion then occurs lifting debris and water many feet in the air. Later in the war records show that the sub was I-45 and on the same day a second enemy sub was sunk in the same area.

On 29 October we (*Intrepid* in 38.2) hit Manila airfields as the Japanese are moving additional planes there to harass our forces at Leyte. These planes come up to fight ours going in and one hell of a dog fight ensues. We lose 11 planes and 16 air crewmen. *Intrepid* launched 112 sorties this day hitting targets primarily on and around Clark Field near Manila. 36 tons of bombs were dropped around Clark Field alone. 40 enemy planes were shot down by Air Group 18's pilots (and six probables) and at least 7 others destroyed on the ground. Many other enemy planes were damaged. The fliers had a good day, but not their home ship.

Shortly after the noon hour on this day a kamikaze (a Val), hit and on fire, singled out *Intrepid* and crashed into tub #10, a 20 mm gun mount (six guns), port side, amidships, killing six men instantly and injuring ten. Of these latter, four would die before the day was out. *Intrepid* was the first of the fleet's fast carriers (CVs)

A shot up fighter lands on 29 October, 1944 without a tail and with extensive damage to left wing and fuselage. (National Archives Photo)

to feel the sting of a suicide plane. Steward's mates were manning this tub and no man left his gun as the Japanese plane came at them. The plane hit the battery of 20 mm guns. Gasoline from the kamikaze's ruptured fuel tanks immediately spread fire all over the area, including the ready rooms (magazine storage) and the catwalks close by. Fire fighters very quickly got the fire knocked down; it took only four minutes to put the fire out. The flames and heat were so intense for that short period of time the men did not have a chance. In the area at the time was Lt. (later Lt. Cmdr.) Charles Devens and Lt. Cmdr. L.L. Hogan. Both took charge immediately. Devens at the time was landing officer and ran to the tub as soon as the plane headed to the tub. The ammo in the ready room began to explode and go off in all directions. The plane, after setting down in the tub on the 20 mm guns, then fell over the side. One man was decapitated and four others killed instantly with another dying several hours later. Six steward's mates were injured and suffered second degree burns. Devens pulled men free of the area and so did Al Brouseau, S1C, who went into the tub and pulled out a man on fire and laid on him, extinguishing the flames. Other men went into the burning gun battery and pulled or helped men out of the fire. One was a radio man, one a pharmacist's mate; all were heroes and thought not of their own safety. Private First Class Bell, a Marine, went to one gun whose magazine was spitting out 20 mm shells. He yanked the magazine off the 20 mm and threw it over the side. Luckily, no shells hit him. The Fire Marshall was Lt. D. DiMarzo who took charge of the fire and his men quickly put out the flames. *Intrepid* repaired her own damage and stayed on line.

Lt. K. C. Crasaw's plane is stopped by Intrepid's crash barrier 29 October, 1944. Pilot is hurt as the plane's prop dug into flight deck. (National Archives Photo)

The Intrepid, seen from the U.S.S. Cabot (CVE 28), as it is hit by a Japanese suicide plane on her flight deck 29 October, 1944. (National Archives Photo)

The suicide plane hits Intrepid's tube #10, port side, explodes and falls into the sea 29 October, 1944. Eight men are killed, two others lost over the side and eight are injured. (National Archives Photo)

Fire produced when Jap plane crashed in Tub #10. Note the scattered plane parts on the deck. (National Archives Photo)

Injured man on deck after plane crashed into Tub #10. (National Archives Photo)

Putting the fires out in Tub #10 after a Japanese plane crashes and falls into the water. (National Archives Photo)

Tub #10 is inspected after Jap plane crashed and the fires put out. (National Archives Photo)

Repairs begin on Tub #10 after attack and crash of Japanese plane. (National Archives Photo)

Burial services on Intrepid for ten Stewards Mates killed in Tub #10. (National Archives Photo)

The last of Tub #10 dead is lowered into the sea as night begins to fall. (National Archives Photo)

Next day, 30 October, we (*Intrepid*) stand out at Samar while Task Group 38.4 (Davison) took over the Leyte cover and we worked a little more to get *Intrepid* shipshape again. Carrier *Franklin* (CV-13) is now hit. Five kamikazes go after her. They get by the CAP screen by coming in real high at 18,000 and 19,000 feet. Radar does pick them up at this time and three are shot down. One manages to get in and he crashes on *Franklin's* flight deck. The *Franklin* is now afire and has a 40 foot hole in her flight deck and #3 elevator is twisted and out of service. The suicider has a bomb in his under-clamps and the explosion destroys 33 planes, kills 56 men and injures badly 60 more.

At the same time the light carrier *Belleau Wood* (CVL-24) is hit by the fifth attacker (he dropped his bomb into the water) and the carrier's flight deck is holed. She loses 12 planes and 92 men of the crew are killed and 54 others badly injured. Her fires are brought under control but it takes several hours. Both carriers with an escort then head for Ulithi and eventually the west coast of the U.S. where they will be repaired.

The ships of Task Group 38.4 are now ready for supplies and refueling and so the Group retired, with their two badly damaged carriers, to Ulithi on 30 October (joining Sherman's 38.3 and McCain's 38.1) leaving Bogan's four-carrier 38.2 (including *Intrepid*) on station alone off the Philippines. Vice Admiral McCain relieved Vice Admiral Mitscher as Task Force 38's carrier boss during their short stay at Ulithi (Rear Admiral Al Montgomery took command of McCain's Group 38.1). Halsey remained in overall command of the Third Fleet.

Task Group 38.2 (*Intrepid*) retired off Leyte on 31 October, just beyond enemy strike range, to carry out sea routine duties. Although most of T.F.38 was at Ulithi, much fire power was left at Leyte to cover the area should the enemy start to move again. The big guns left at Leyte were the battleships *California* and *Mississippi*, the cruisers *Nashville*, *Boise*, *Phoenix* and H.M.A.S. *Shropshire* and a dozen or so destroyers. This group left at Leyte were under command of rear Admiral Weyler.

The Japanese at this time have been moving planes into the area and on 1 November at about 0900 hours some of them are air borne and they head for Leyte Gulf. The first enemy wave of planes are torpedo bombers and as usual they fly low, close to the water. Some dive bombers get through the screen and then they went high into the clouds and they dive down at steep angles toward the ships below. Anti-aircraft fire was heavy, but some got through. One dove on the destroyer *Claxton* doing much damage. Another one, a Betty bomber (twin engine), already on fire from being hit, crashed into the bridge of destroyer *Ammen* and continued on into the water. Her guns did not stop firing as she shot down a second one heading for her. The *Ammen* had five killed and 21 injured from the first kamikaze hit. Destroyer *Bush* was attacked about 0930 hours by torpedo bombers and other bombers which dropped a series of bombs on her while fighters strafed her decks. The *Bush* outmaneuvered it all and was not hit. The *Bush* is attacked at least a half dozen times in the next two hours and she is not damaged.

At this time destroyer *Killen* has a half dozen planes come at her all at once. She downs four of them, but one drops a bomb on her, forward, killing 15 men instantly and setting her on fire. She also begins to take on water forward. At about 1130 hours Admiral Weyler begins to shift his destroyer screen around and to put some of them around the battleships and cruisers. Four destroyers arrive at this time and they check in with Admiral Weyler. At about 1300 hours destroyer *Abner Reed* is attacked by two Vals. One is shot down and the other is hit and afire but continues on, crashing into *Abner Reed.* A fire is ongoing and the flames get to a magazine and a terrific explosion takes place. She is now taking on water and is developing a bad list. The order is given to abandon ship as she now has a 30 degree list and is settling in the water. Just about the whole crew gets off her and at 1400 hours the *Abner Reed* rolls over and goes down by the stern. 22 men were lost. The enemy air attacks continued all day until darkness began and in one other attack destroyer *Anderson* took a small size bomb in the area of the bridge and she lost some men.

Things were getting hot around Leyte so two of the Task Groups at Ulithi, 38.1 and 38.3, were ordered to return to action. The battered 38.4 (Davison) remained behind at Ulithi. All three Groups were off Luzon on 2 November. The Army Air Force as yet has not come into the air war here as the field at Tacloban is not ready to land them. Halsey is worried about what may still come from the enemy and he contacts Nimitz at Pearl Harbor. They agree that at least three carrier groups concentrate on this area and to try to hit the kamikaze take off points and both agree that the carriers must be protected from the kamikaze planes.

It is early on 2 November. Admiral Bogan's unit (*Intrepid*) is ordered to send up a strike and search mission for a reported enemy naval force on the move west of the Philippines. No enemy force was located but they did discover another Japanese force in Ormac Bay landing reinforcements there and also many enemy planes on the ground at enemy airfields on Luzon. Halsey regroups all carrier forces and plans are made for massive strikes on the enemy areas to begin 5 November. The main targets are Japanese fields to knock out all these planes that they have ferried in. It is to be a hunt, kill and demolish series of attacks to wipe out the Japanese air force in the whole area. Because the Army air is still not able to cover MacArthur, the Third Fleet must do so. Nimitz asks MacArthur to release the third Fleet as soon as he can so they can join Carrier Task Force 38. The Task Force is refueled on 3 November and late at night the force heads north to Luzon. At about 2330 hours the cruiser *Reno* is torpedoed and has to drop out of the formation and retire for repairs. Two men were killed.

A SB2C swings into the catwalk and burst into flames as it attempts a take-off at dawn 3 November, 1944. (National Archives Photo)

SB2C #53 falls into catwalk forward starboard bow as he is being spotted forward. (National Archives Photo)

At 0600 hours on 5 November the Task Force is in position off Luzon and begins launching aircraft. 38.1 will hit northern Luzon with Clark Field and Lingayen Gulf as her targets. 38.2 (*Intrepid*) will hit southern Luzon, Clark Field, the Verdi Island passage and will hit Mindanao airfields and cover the north Sibuyan Sea area. 38.3 will hit the Manila Bay area. Admiral McCain keeps the fleet at least 80 miles off shore and has a large number of planes up (CAP) to cover the ships should the kamikaze planes show up. The strikes are made and the enemy must have been asleep. Because the enemy was slow to react to the raids on their bases, only a few planes got airborne and many more on the ground were hit before they could get up. The strikes last two days over the enemy fields in which many planes had been hidden and camouflaged at the fields' side and some were hidden in tree areas. It is reported that 439 enemy planes were destroyed (most on the ground) in the two-day attack by Task Force 38.

Intrepid planes made 194 sorties in these two days, dropping some 33 tons of bombs. They claimed 57 enemy planes destroyed and 33 damaged. Air Group 18 lost one fighter and one bomber, the fighter to a flak hit which killed the pilot. The single most important hit at Manila Bay by 38.3 was the sinking of the heavy cruiser *Nachi* which was Admiral Shima's flag ship at the battle at Surigao Strait. We did not get away clean in these strikes as the Task Force as a whole lost 25 aircraft due to aerial fights with the enemy. Of these 25, also lost or missing are 18 air crewmen. Also, the carrier *Lexington* is hit by a suicide plane, one of four which attacked her. The plane hit outboard of the island (starboard side) knocking out some gun controls and killed 50 men with another 132 injured. *Lexington* put out her fires quickly and stayed on line and continued flight operations.

Late in the day of 6 November all ships pull back for fueling, putting us some 400 miles to the east of Manila. 38.2, including our *Intrepid*, escorted *Lexington* back to Ulithi, leaving the Task Force on 7 November. At this time Task Group 38.4 would leave Ulithi to take our place on line. The strikes at Luzon did as we had hoped: It reduced the plane attacks on our ground troops on

On 7 November 1944, another downed pilot comes home by way of the destroyer that fished him out of the water. We paid 20 gallons of ice cream and a case of oranges for his return. (National Archives Photo)

Leyte. It is now the rainy season in this area and it did rain: Some 35 inches of rain. The month of November was so wet that the work on the new airfields was continually held up because the fields were flooded and bogged down in mud. Tacloban finally was finished enough so the Army was able to bring in some planes to take some of the load off Task Force 38 carriers. Three or four other fields under construction were held up due to the rains and the flooding. Another new field on Leyte finally became operational on 16 December and on 18 December a field on Guiuan Island off Samar was operational. In the last 10 days of December 1944 both Army and Navy planes were now coming in and landing on these fields and would take on some of the work that the carriers had been doing for so long a time.

The enemy now began to come out at night to harass the Task Force. As soon as the Army and Navy planes retire for the night, the Japanese go up and everyone is at General Quarters or T.D. Now it seems that the Tokyo Express is back also as the Japanese try night landings of men and materials on their scattered outposts and beaches on Leyte. They go all out to get American troops off the island. By 12 November the Japanese had landed 45,000 troops and 10,000 tons of material on the west side of Leyte. When the U.S. force went up against Leyte in the invasion on 20 October, the Japanese force was only 20,000. To remedy this Tokyo Express problem we must have night patrols in the skies to stop the attempted landings of the enemy under cover of darkness. The Marine squadrons in the Solomons were the ones to do just that. Admiral Halsey makes contact with General MacArthur and that is what he suggests we do: Bring in the Marine air squadrons to fly night missions in the area.

We now have Army B-25s and heavy B-24s as well as P-38 fighters on the new fields in and around Leyte. The Japanese are moving men and materials from the Visayan Islands group and from Cebu to Ormac on Leyte. On 5 November and the 7th and again on the 8th, the planes take off and bomb those areas. The bombers are covered by fighters from the new Tacloban field. The next night the Japanese try again. They have some 2,000 troops and some 6600 tons of supplies and they try to get to Ormac on Leyte's west coast. The B-25s crush the attempt. The Tokyo Express is being slowed down.

We now also have the 13th Army Air Force based on a new field on Morotai. They arrived about 1 November. On 10 November the Japanese gather ships to make another attempt to land troops in Ormac Bay. This convoy is destroyer escorted and has some 10,000 troops on five transports. Task force 38 is some 400 miles away, getting ready to refuel. A search plane in the area spots this convoy and sends a message to Admiral Halsey, giving him the location and description of the convoy. MacArthur, hearing of this force, sends an urgent message to Halsey asking that T.F. 38 cancel their fueling rendezvous and head at best speed to the area and launch aircraft against this force. Halsey instructs Admiral Sherman who is in temporary command of T.F. 38 at this time to cancel fueling, reverse course and proceed at top speed west to attack the troop transports. 38.1, .2 and .3 head west (*Intrepid* and 38.2 are at Ulithi at this time).

It is now 0600 hours on 11 November and the three carrier groups of Task Force 38 are within 200 miles straight off San Bernardino Strait and the launch of search aircraft begins while the carriers continue to close the distance. The Japanese convoy is spotted approaching Ormac Bay in the Camotes Sea. The sighting report goes back to the carriers who are at this very moment launching planes. In no more than 45 minutes 347 planes are headed to Ormac Bay. The lead transport is hit first and sunk. (One report states that the enemy convoy had only three transports, another states five). With the convoy now only one mile from the beach at Ormac the planes attack and sink all the remaining transports. Hundreds and hundreds of enemy troops are in the water and just about all are drowned.

The Eddie Peabody Group entertains the crew at Ulithi. (National Archives Photo)

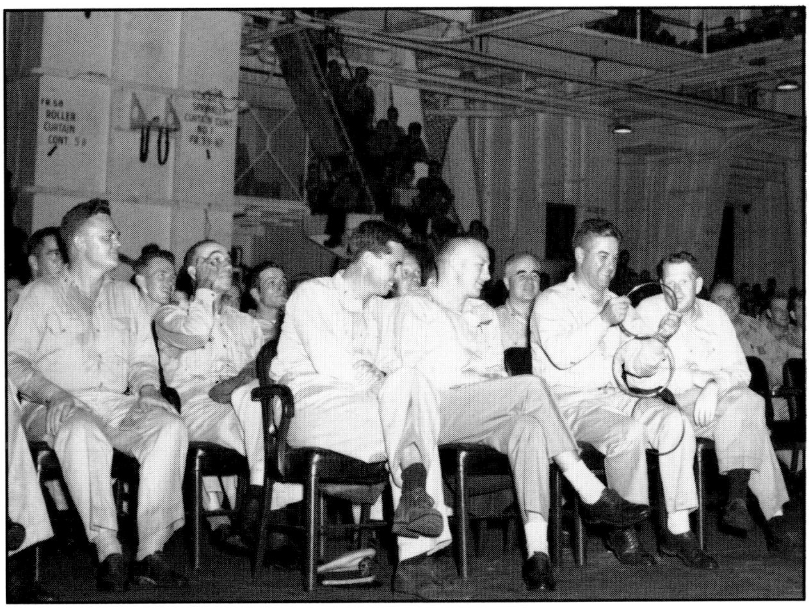

Officers play with old ring trick as the Eddie Peabody Group entertains the crew. (National Archives Photo)

The attacking force was in two waves. When the second wave began their attack the enemy had gotten some planes in the air and they faced the U.S. planes coming on target. The enemy planes numbered between 25 and 35. 16 are shot down. Also hit are the four destroyers and they are all sunk. The record shows that the four destroyers sunk in this area were sunk in the same attack at the same time. They were the *Hamanami, Naganami, Shimakaze* and the *Wakatsuki*. All are listed as sunk by aircraft. The stragglers of this convoy, destroyers *Akebono* and *Akishimo* are also sunk in the bay by the carrier planes shortly afterward on the 13th. The carriers lost nine aircraft in this action.

In two weeks the Japanese will attempt to make another landing. General Yamashita is now behind these "Tokyo Express" convoys. He has meetings with the

New aircraft are loaded unto the Intrepid at Ulithi Lagoon. (National Archives Photo)

field marshall and southern army commander, Terauchi. The meeting was held at his Manila headquarters. In short, his (Yamashita) wishes were that all attempts be stopped to regain control of Leyte. But Terauchi would not buy it. He wanted the reinforcement attempts to continue. Terauchi wanted everything done that could be done to regain control of Leyte. His orders to continue brought needless destruction and thousands of more deaths to the Japanese people.

A new dive bomber is safely on board at Ulithi on 12 November, 1944. (National Archives Photo)

Admiral Kurita has now re-grouped his ships and he is at north Borneo according to reports. Admiral Halsey wants to hit that area next. Admiral Nimitz talks to General MacArthur about Halsey's request. MacArthur would prefer that the Task Force continue to hit Luzon. Halsey is somewhat disappointed when told of MacArthur's reply. Task Force 38 refuels; it is 11 November late in the day and refueling will continue into next morning. Our planes have returned from the convoy strikes. On the 12th of November the fleet moved back to the Luzon area. Japanese scout planes report back on the sighting of Task Force 38 headed west. Two Japanese attacks are repulsed by the carrier planes. We now prepare to hunt down and strike enemy shipping and it isn't long before the search planes report back a sighting of a convoy of transports with 5 destroyers and at least one cruiser on the move from Manila Bay. The carriers send up a large strike force to hit this convoy. Again all prime ships are sunk in Manila Bay. The sunken ships were the light cruiser *Kiso* and the destroyers sunk were *Okinami*, *Akebono*, *Hatsuharu* and *Akishimo*. All transports were also sunk on this 13th of November at Manila by the planes of Task Force 38.

A large group of enemy planes came out to protect this convoy. When it was all over an estimated 84 enemy aircraft were destroyed. Task Force 38 lost a total of 25 planes due to ground fire and from the ships which were attacked. Task Group 38.2 (*Intrepid*) did not take part in this strike in Manila Bay as we were at Ulithi getting fuel, supplies and some service work to the ship. We now leave Ulithi (on 14 November with carriers *Hancock*, *Independence* and *Cabot*) and at the same time task Force 38 pulls back eastward for fuel. Admiral Sherman's Task Group 38.3 continues on to Ulithi as Task Group 38.2 (*Intrepid*) rejoins Task Force 38. From

Captain Giles E. Short and Commander W. E. Ellis during a Captain's inspection on the Intrepid. (National Archives Photo)

the 15th to the 18th *Intrepid* refuels destroyers, has gun practice and is in general sea routine.

On 19 November Groups 1, 2 and 4 make Luzon strikes hitting key installations and the airfields. One of these strikes smashed over 100 enemy planes in the air and on the ground, *Intrepid's* Air Group 18 pilots claiming 20 of these. Task Groups 1, 2 and 4 lost a total of 13 planes on this strike. *Intrepid* sent off 273 sorties on this 19 November, losing two fighters and one pilot to ground fire at the target.

Now comes the enemy to hit Task Force 38. Enemy raids are made on us, sometimes one plane and at other times 5, 6, or 8 planes come at us. The Japanese try no less than 10 times to hit us and they are shot down. The only near miss was on the destroyer *Collette*. She was attacked by Betty bombers and these planes sent two torpedoes at her. The *Collette* shot the Bettys down and out-maneuvered the torpedoes The 19 November enemy raid saw much anti-aircraft fire but no damage to Task Force 38, except for the trade of 13 U.S. aircraft against over 100 enemy planes. We pull back and take it easy on 20 November. Some ships top off on fuel.

Back on 25 October we had planned a strike on Yap Island but due to other actions the strike was put off. On 22 November Admiral Davison is sent to Yap with Task Group 38.4 to hit it and burn it. 96 fighters make the run to Yap carrying rockets and Napalm bombs. It is now we find that Napalm bombs don't do what they are supposed to do, that is, ignite. More than 50% of the bombs refused to burst into flames. Needless to say this weapon is undependable and must receive more experimental work on it to make it work. (Improvements on the bomb were made and at Okinawa it worked fine and in later years, it was the key weapon at Korea). Group 38.4 returns to the Task Force and then both Groups 38.3 and 38.4 head back to Ulithi for services and provisions as well as ammo. Only two Groups are now standing out at the Philippine area, Task Groups 1 and 2. Carriers available are four *Essex* class CV carriers (like *Intrepid*) and three light, *Independence* class CVL carriers. At least 16 carriers usually made up Task Force 38

These ships move out on the night of 24 November and on 25 November Bogan (38.2) and Sherman (38.1) are at a point 165 miles east of Manila and we launch planes on a strike to hit Luzon early on 25 November. In a search over Lingayen Gulf the *Ticonderoga* planes spot the heavy cruiser *Kumano*. This cruiser was hit bad in the battle off Samar and again hit by one of our subs soon after. She is now sunk by *Ticonderoga* planes in Dasol Bay where the water is not deep and the cruiser is not fully submerged. At the same time a small convoy is located about 15 miles away to the southwest and the convoy is sunk. The carrier planes find another convoy some 30 miles north of Lingayen Gulf and two small transports are sunk there.

On this day of 25 November the Japanese have many planes in the air just as we do. Over Luzon we shoot down 26 of them and hit another 29 on the ground. The enemy now has had enough as they come after the carriers, not only their planes but their subs move in on Task Force 38. It seems that they now have our number. We are at half force as Groups 3 and 4 are at Ulithi. Our CAP is not as good as usual and now our position is known. It is noon of 25 November and *Intrepid* is some 60 miles off Luzon. Task Group 1 is some 35 miles away. We both have our CAPs in the air. We are preparing to launch our third strike at this time as our second strike is landing. Bogeys approaching at 25 to 30 miles, the PA announces. The planes are coming from the direction of southern Luzon. Radar says many — so many that the radar face is covered with blips.

The last plane lands, although plane is damaged and both pilot and gunner are injured. (National Archives Photo)

Hancock, *Intrepid* and *Cabot* had started to launch planes at 1226 hours. It is now 1229 and our Admiral Bogan on *Intrepid* orders full speed. Our CAP goes out to meet the enemy. Six enemy Zekes are attacked by CAP and they shoot down two and hit a third. The others dived low and continued on. It is 1234 hours and one of the Zekes goes for the *Hancock*. He is hit and blows up no more than 300 feet from *Hancock*. Sections of the plane's wing fall on the carrier's deck, starting a minor fire and knocking out one 20 mm gun. Radar shows many bogeys on the screen. It is 1248 hours and 12 planes get by our CAP flying only 100 feet above the water. All guns open fire. The 5/38s, the 40 mms, and all 90 20 mms at different times when the position of the ship allowed them to train on a target. It is 1253 and *Intrepid* takes a suicide plane. The plane penetrates the flight deck down onto the hanger deck. One 20 mm gun is torn lose by the plane as he hit the ship. The plane's bomb explodes on the hanger deck which forces the flight deck upwards. We have serious fires ongoing and damage control goes to work on the fires.

The ship continues to repulse the attacking planes. We are sending up much heavy black smoke from our fires. In a couple of directions we can see nothing at all. The carrier *Cabot* is now hit by a suicide plane, forward, port side edge, the plane landing partly on the catwalk and partly on the deck. In just two minutes a third kamikaze heads for *Cabot;* he just misses and goes into the sea. The *Cabot* has two open six foot holes in her flight deck, another small hole in her hull, the catapult is damaged and 36 men are killed and 16 injured. One hour later she recovers her own planes.

Sherman's Group 3 is also under attack. Two suiciders aim for *Essex*. It is 1255 hours and the two planes head to the ship and one is shot down. The other is also hit but continues on; he too hits forward, portside, just as the *Cabot* was hit and like the *Cabot* damage is not bad. The *Essex* lost 15 men. The *Hancock* was also hit at 1234 hours. More planes come on at 1248 hours. At 1259 *Intrepid* is hit again, only six minutes after the first kamikaze hit us. This time the plane came in very low from straight astern of the ship. He was in perfect position to land on the deck. At most, he was 160 feet above the water. Not too many guns can train on him. The only guns that could line up on him were aft and on the fantail. His nose dropped and he hit at a very shallow angle. He too went through the flight deck and he too was carrying a bomb. This bomb also exploded (like the first) on the hanger deck and more explosions and fires erupted. The plane exploded into hundred of pieces all over the deck and it's engine slid forward and the pilot's body also slid forward some 60 feet. The body had its left leg torn off and the pilot's head on the left side was blown away. This second crash tore out some of the arresting gear and now we cannot land planes even if we repair the upheaved and holed flight deck. Two of the ship's elevators were damaged and every plane on board was damaged except one which was later flown off.

Intrepid turns right so as to put more guns on the incoming planes 25 November, 1944. (National Archives Photo)

Ships under attack, as seen from the Intrepid. (National Archives Photo)

Intrepid turns hard right, as plane turns for attack. (National Archives Photo)

Intrepid is hit by one kamikaze, as another plane starts the turn for a successful attack on the ship, as seen from the USS Iowa. (National Archives Photo)

Gas flames and molten metal fly through the air as the Japanese plane crashed into the Intrepid. (National Archives Photo)

Japanese planes fill the sky around the Intrepid, as one explodes aft of No. 2 elevator. Smoke is choking those stationed on the island. (National Archives Photo)

Jap Zeke is hit, but he still comes on and crashes the Intrepid. (National Archives Photo)

First kamikaze wreckage is on deck with fires about extinguished. Minutes later a second suicide plane hits with a 500 lb. bomb. (National Archives Photo)

Remains of first suicide plane on deck with pilot's body in wreckage. (National Archives Photo)

Japanese plane with 500 lb. bomb has gone down through the flight deck to the hanger deck. (National Archives Photo)

Fires burn up through the flight deck started by the crash of a suicide Japanese plane. (National Archives Photo)

Plane parts are scattered over the flight deck, while fires in hanger deck continue. (National Archives Photo)

Gas fire is red hot in wreckage of first suicide plane to hit the deck of the Intrepid. The pilots body was found amongst the rubble. (National Archives Photo)

Intrepid afire after attack from kamikazes. The New Jersey can be seen in foreground. (National Archives Photo)

The men desperately try to extinguish the raging fires on the Intrepid. (National Archives Photo)

The Japanese plane blew into thousands of pieces as it hit the flight deck. (National Archives Photo)

Men covered in oil try to put out the fires in the remains of the hanger deck. (National Archives Photo)

Men begin to salvage prop from blown up F6F fighter on hanger deck. (National Archives Photo)

Intrepid planes return from strike to find they are unable to land due to the damage caused by the kamikaze attacks. (National Archives Photo)

The wounded are treated on the flight deck. (National Archives Photo)

The damaged sections of the flight deck are removed and replaced with plates so that the planes will be able to land. (National Archives Photo)

The fires are out and the men inspect the damage to flight deck from Japanese bomb which penetrated the flight deck and exploded on hanger deck. Anthony Zollo in man in summer tans with life belt around waist walking towards hole to the left of the photo. (National Archives Photo)

Ship fitters and repair parties work on damaged flight deck. (National Archives Photo)

The deck is cleared leaving only the remains of the kamikaze plane's motor and the pilot. (National Archives Photo)

Burial ceremonies for the many who lost their lives. The Marine to the right of the flag is Tony Zollo. (National Archives Photo)

While anchored in Ulithi Lagoon 30 November, 1944, Admiral Halsey escorted by Captain Bolger, Lt. Cmdr. Thompson (Hull Department), and Lt. Cmdr. Walace (Damage Control) inspects the damaged inflicted by the kamikazes and their bombs. (National Archives Photo)

The burned out hanger deck is inspected by Admiral Halsey. (National Archives Photo)

Vice Admiral McCain and Captain Bolger inspect bulkhead frame twist. (National Archives Photo)

Vice Admiral John S. McCain, along with Lt. Cmdr Walace, inspects the damaged Intrepid. (National Archives Photo)

One of six decorated Steward Mates of Tub #10 is presented an award by Captain Joe Bolger. (National Archives Photo)

Captain Joe Bolger presents awards to crew members on the fire-damaged hanger deck on 11 December, 1944. (National Archives Photo)

The fires continued to burn. We had below armed and gassed-up planes ready for the next strike. They blew up and added to our fires and our damage. We had some 75 planes in the air at the time and other carriers took them on board, fueled them, armed them and the planes flew to Tacloban and landed there. A strike was leaving Tacloban that afternoon and our planes went along on the strike. *Intrepid* remained on line during the attack . We lost 69 men of the crew and had 85 injured seriously in the two suicide crashes. During these attacks our pilots and those of the other three carriers in our group shot down 19 enemy planes and 13 more were blown out of the sky by the ships' guns. These 32 enemy planes were destroyed in 20 minutes.

By early evening we have the fires out and the crews have already more or less cleaned up the flight deck. All but the suicide plane's motor and the Japanese pilot's body have been pushed over the side. Some timbers here and there still smolder with little puffs of smoke coming from them. The welding and burning tanks are already on the flight deck and the men burn away bent and twisted metal. The upheaved flight deck area is being cut away so we can lay down sheets of steel and plywood to cover the holes. Our damage is extensive an I do not believe that Ulithi can service us; we will have to go stateside. Perhaps we will have Christmas at home this year. If we go stateside we should be at Hunters Point Yard at San Francisco by 20 December. I know that plans were made to strike the Visayas area tomorrow, 26 November. I guess that will be called off now.

Congratulations For Service On 25 November, 1944

Message from Commander of the Third Fleet to all ships 25, November, 1944:

"The Nips scored some points today, but they lost the game. Well done on the overall alert and aggressive performance of the Task Force and better luck with the interception next time."

Admiral Halsey

Message from Commander of Task Force 38 to Commander of Task Group 38.2 (Carrier Intrepid):

"Both Intrepid and Cabot handled a tough situation with cool resolution."

Admiral Halsey

Message from Commander of Task Group 38.2 (Rear Admiral G. F. Bogan in Intrepid) to the men of the Intrepid:

"The performance of the officers and men of the Intrepid in action under the most trying conditions while fighting fires was an inspiration. The loss of our shipmates is deeply regretted. You have the admiration of the whole Task Group and its commanders."

Rear Admiral Bogan

Message from Commander of the Third Fleet to the Intrepid and Rear Admiral Bogan:

"The normal and efficient performance of C.T.G. 38.2 and her flagship Intrepid, today in spite of enveloping flames and smoke was an inspiring sight. Well done!"

Admiral Halsey

The Marines sunbathe on No. 2 elevator at Ulithi, our R & R stop over and repair base in the Western Marianas Islands. (National Archives Photo)

We prepare to leave the Philippine area and all of Task Group 38.2 is coming with us. We leave Task Group 1 alone at sea in the Philippine area. I would believe that other Task Force 38 forces will now be recalled from Ulithi to return here. The *Independence* CAP is overhead looking out for Task Group 2. Admiral Halsey as once before, is now concerned about the kamikaze attacks. I would suppose that he sends out strikes to wipe them and their fields off the face of the earth.

I had the duty the afternoon of the 25th as orderly to Admiral Bogan and when General Quarters sounded I went to my 20 mm, two decks up in the Island outboard on the starboard side. Most of the starboard guns did not get to fire as the planes came in from the port side and from aft — except one plane that came in on the starboard beam and we blew him away at about 1,000 yards. When the action was over I went down to the flight deck to see for myself what our damage was. Where my gun is located I have the Island bulkhead behind me and I can not see the flight deck area except way forward and way aft. The only way to get to my gun is through the Island. The irony of all this is that this was to be our last day of baby sitting the Philippines as we now had a base with three fields operational in the Leyte area. The Army was now landing planes there and they would cover the Philippine area more or less We actually spent four weeks, plus, here waiting for those airfields to become operational. I may be wrong but I figure we have been on the attack against the Japanese for some 84 days without let up except for two days at Ulithi.

Admiral Halsey now makes a statement: "*Further casual strikes did not appear profitable, only strikes in great force for valuable stakes or at vital times would justify exposure of the fast carriers to suicidal attacks — at least until better defensive techniques were perfected.*" The ships damaged in these 84 days were the *Cabot*, *Intrepid* (twice), *Lexington, Franklin, Essex* and *Belleau Wood.* The

Princeton was first hit and there was the action by the Taffys in which we lost the escort carriers *Gambier Bay* and the *St. Lô.* Also lost on 25 October were the destroyers *Johnson* and the *Hoel.* The destroyer escort *Samuel B. Roberts* was also lost at the battle off Samar. In early November Admiral Halsey tried to get the Marine air wings to take over the kamikaze plane situation. He wanted them to be based possibly at Tacloban, but nothing ever came of his request. Now, as November ends, it is MacArthur who makes the same request. I have more than once thought to myself: Why don't the carriers have steel covered flight decks? So much damage below and many fires could be eliminated with steel decks. Interestingly, the British carriers have steel reinforced flight decks.

Home Again

From my diary beginning 26 November:

November 26. Our position at 2000 hours: Lat 14°15'N, Long. 133°05'E. We were on the gun at twelve noon yesterday and right through the night. We got off the gun this morning at 0720 hours. Everyone goes on an inspection tour to see the damage to the ship. As I said earlier, we are in bad shape. We have no radio or radar. We have burial ceremonies at 1300 hours for our dead. All Marines doing the honors. We put about for Ulithi Island. We have Torpedo Defense at 1730 hours. We secure.

Burial services for our dead takes place in the Philippine Sea with full military honors, by the crew, officers and the Marine detachment. I am part of this detail.

November 27. Our position today: Lat. 10°00'N, Long. 139°56'E. We are 25 miles from Ulithi. We go to Torpedo Defense at 1500 hours, our 5/38s are firing and no one knows why. Land is sighted at 1600 hours. We drop anchor at Ulithi at 1730 hours.

November 28. Port routine. General Quarters at 0600 hours. Torpedo Defense at 0800 hours. We transfer our flag and staff. The ship is being stripped of guns, gunsights, radar parts, antennas, etc. Our skipper has received the Navy Cross and with gold star, in lieu.

November 29. Port routine. Admiral Bill Halsey comes on board to inspect our condition. He orders the ship state-side for repairs. It's a little too much work to be done at our yard at Ulithi.

November 30. We take on board some passengers who will go back with us. Vice Admiral John C. McCain comes on board the *Intrepid* to have a look, to inspect the ship's damages. We are at Ulithi lagoon.

December 1. We get ready to move out and head for Eniwetok.

December 2. Under way at 1630 hours.

Our position today at 2000 hours: Lat. 10°12'N, Long. 140°74'E. We moved clocks ahead one hour last night at 2115 hours.

December 3-5. Sea routine. We go to Torpedo Defense at 1800 hours, on 3 December.

December 6. We arrive at Eniwetok at 1730 hours.

December 7. *Intrepid* ups anchor and departs Eniwetok at 1400 hours. We head for Pearl Harbor. This is Pearl Harbor Day as we leave Eniwetok lagoon.

December 8-12. Sea routine.

December 13. *Intrepid* at Pearl Harbor. Captain Joseph W. Bolger presents awards on board *Intrepid.*

December 14 & 15. Port routine.

December 16. Left Pearl Harbor about 0800 hours for Alameda, California.

The passage north eastward to San Francisco was uneventful. *Intrepid* and her escorting destroyers passed the Farallon Islands, 30 miles off the Golden Gate Bridge, about 1215 on 20 December, and by 1600 hours she was tied up at Alameda Naval Air Station. Repairs began almost immediately on the battered carrier. At Alameda I left the ship for a 21 day leave. A lot of guys are putting in for a transfer as they don't want to hang around here for three months or more, so I also put in for a transfer before I leave the ship. I head across country by train. I have Christmas at home for a change. In January I am examined and tested and found fit and qualified for promotion to Corporal. I do not know of this promotion until I read it in my service record in February 1991.

Christmas of 1944 at Hunter's Point dry-dock in San Francisco with the WAVES onboard. (National Archives Photo)

I returned to the ship, now at Hunters Point, California, on 4 January 1945. I go aboard and check in. It is then I am told I am transferred to the ammo dump at Hingham, Massachusetts, fifteen miles from home. At first I don't believe it. I stay on board just long enough to pack my Sea Bag and I'm off to Boston again. I make my 6th trip across the country by train. No one tells me I have received another stripe and there is no mention of it in my transfer papers and I do not know of this promotion until some 47 years later. U.S.N.A.D. at Hingham was not notified either and I am discharged ten months later with the old rating I had held before my transfer. It looks like Uncle Sam owes me the paper work and discharge listing the new rank and also about eleven months' pay increase.

The War Continues

The war did not stop of course while *Intrepid* was stateside undergoing repairs of her kamikaze battle damage. The notes I continue to make read like this:

General MacArthur decided to make a landing on the west coast of Leyte. He chose the special day of 7 December. Two divisions would land on the beach near Ormac Bay. As he made the first landing the suicide planes came at him like the leaves blowing in a stiff wind in late Fall. They were everywhere. Right off, the destroyer *Mahon* and the destroyer *Ward* were hit and sunk. Some damage is done to the fleet but no matter what they tried or did, they could not stop or even hinder our operation, as at this time our fleet had grown to a tremendous size. The suicide pilots are reported to number about 50 at this time and a program goes into effect to recruit more, but it is too late. At this time the Army is fairly active in the skies and so too is the Marine air group which finally got to the area. Before this new landing at Ormac Bay, the waters had to be swept by the mine sweepers. Because of the shallow waters no deep draft ships took part.

The cover for the landings was a dozen destroyers and the landing craft were 12 LCIs and 12 LSMs along with four LSTs. It is about 0630 hours as the destroyers open and bombard the shore areas where the new landing will take place. At 0715 hours the troops hit the beach and all goes well as by 0900 hours five waves had gone ashore including much unloading of supplies and material. All this action is on the west coast of Leyte and at an area to the south known as Deposito. By night fall the U.S. troops had control of an area over one mile inland by some two miles along the shore of Ormac.

Word now comes in of an attempt by the enemy to land forces north of Ormac, arriving by way of the Visayan Sea, some 65 miles to the north and headed for San Isidro Bay. The Tokyo express is now again in operation. The Marine air squadrons are called in from their field at Tacloban and they hit this enemy force which now is within 18 miles off Ormac. Four large transports and one small ship are sunk by the Marine fighters but not before some 4000 enemy troops get ashore. Time now for the enemy air to begin to harass this Ormac operation and that is exactly what happens. Radar picks up a large blip on the screen; it is 0940 hours on this 7 December and 12 enemy bombers are approaching the area. The P-38s take off with some other fighters to intercept the enemy force flying in. The U.S. destroyers off Leyte near Ponson Island came under attack by these enemy planes. The enemy drops torpedoes against the destroyers and when their torpedoes made no hits on the destroyers, the enemy dived into the U.S. ships. The P-38s knocked down three enemy fighters and hit two of the bombers. The two hit bombers begin a suicide dive on the destroyers and they are followed by others. A total of nine enemy planes head for the destroyers to crash on them. All are shot down and miss the destroyers except two, which are hit but continue on. These two head for the *Ward* and the *Mahan*. The *Ward* is not hit at this time but the *Mahan* takes two hits. The *Mahan* is afire but still full underway. This action began about 0930 hours. The fires on *Mahan* are now uncontrollable and she stops dead in the water. The abandon ship order is given; it is now 1000 hours. Other destroyers come over to pick up the survivors of the *Mahan*. At 1150 hours the *Mahan* is still afloat and is ordered sunk by U.S. gunfire.

The *Ward* had three planes come at her. The first crashed into her port side opening up her hull and causing an explosion and fire. A second plane and a third made a run on the *Ward*, both being shot down. *Ward* is now burning and efforts are made to knock down the fires; the fire is too great and orders are given to abandon the ship at about 0940 hours. Again, the other destroyers pick up the men in the water and again the *Ward*, like *Mahan*, refused to go down and so it too is sunk by U.S. gun fire. The *Mahan* had 10 killed and 32 injured but not one single life was lost of the *Ward's* crew. Both these destroyers had a long combat history in this war, seeing action at Santa Cruz, Lae, and the Solomons back in 1943, and the landings at Cape Gloucester, Saidor and others. The *Ward* has the distinction of firing the first American shot of World War II: She was the destroyer that fired on and sank one of the five midget subs at the entrance to Pearl Harbor on December 7, 1941.

It was 1005 hours when the *Ward* went down and not one hour later the enemy came back again to hit the remaining destroyer fleet. The destroyers get up steam and they turn for Leyte as the attack begins. In the next four hours there will be four enemy attacks on these destroyers. In this attack the enemy planes pursue the retreating destroyers. The destroyer guns are on the planes and within seconds three are blown apart in mid air. Then one more comes on from straight forward at the *Liddle*, a difficult angle to fire at. Some guns forward are on the attacking suicide plane, but he comes on and straight into the superstructure and he explodes on contact wrecking the bridge area and the combat and radio areas including the captain's sea cabin. The captain is killed as are several officers on the bridge. 36 were killed and 25 injured, some very seriously.

As other ships attempt to assist the *Liddle* another attack begins. This time one of the planes is very high and it is a twin engine bomber. The P-38s are in the air now and over the destroyers. The bomber dives very quickly on one of the destroyers and drops a large bomb. The bomb drops into the water some 50 yards to port as the destroyer's gunners hit the bomber blowing her apart. Another suicider with a torpedo in his rack goes for another destroyer; he is shot down while at the same moment another kamikaze heads to the destroyer *Lamson*. The guns are on it and it is hit but the suicide plane continues on to the *Lamson* which turns quickly and sharply to port as the suicide plane comes in on the ship. Because of the hard turn left the suicide plane's wing hits the destroyer's stack, spinning the plane around and he crashes on the destroyer. A fire erupts and is uncontrollable. The plane wreckage and the bomb it carried jams the hatches in this area, and men below are trapped and burn to death when they cannot open the hatches to get out. 21 are killed and 50 injured. At first it is thought that the destroyer is lost, but she is taken in tow. It did not stop there as at about 1530 hours the kamikazes come back again. This time only one plane and he is shot down.

10 December. Enemy planes made a run on Leyte this day. This time it was eight bombers — twin engine Bettys. The first ship hit was the destroyer *Hughes*. The action was in the Surigao Strait. The *Hughes* is hit amidships and has two dozen casualties. A transport (*William S. Ladd*) is hit and is damaged so badly she later sinks.

11 December. The Tokyo Express is on again for next day. This time they are determined to land troops on the west side of Leyte. Five transports heavily loaded with troops and five destroyers as escorts make the attempt with some 30-odd planes in the air over and around this convoy. It is early morning, about 0815 hours, as the Marine Corsairs take off to hit this convoy. The planes are carrying 1,000 pound bombs. Hits are made on three of the transports and they sink taking most of the troops on board with them. Two enemy planes are shot down. The Marine planes leave the area and some 40 planes both Army and Marine come in to finish the job on the enemy convoy which is nearing Leyte. One more transport is sunk and two others are hit and run upon the beach fully afire. The enemy succeeds in putting some troops ashore.

At the same time a convoy of U.S. ships also heads for the west coast of Leyte with troop reinforcements. This convoy has six destroyers with it as cover. The convoy is attacked by the enemy at 1700 hours. The enemy has ten planes total. Four of them head for the destroyer *Reid.* The first plane hits forward on the starboard side, opening up the hull to the sea. A second one is hit in mid air and explodes; the third misses on a run over her and the fourth comes in low on the water on the port side and plows into the area between her guns aft. The *Reid* capsized and sank, it is said, within two minutes. Less than half of the *Reid's* crew is picked up; the rest are strafed by the enemy planes or went down with the ship. Later on this day the troops of this U.S. convoy were unloaded at Ormac about 2300 hours, the attack on it having taken place at 1700 hours. The attack lasted only ten minutes as the Marine fighters drove the enemy away.

12 December. It is now early morning of 12 December as the U.S. ships turned north to unload more troops some 15 miles away. As the landing craft and its destroyer cover approach the designated area, they find that the enemy is also trying to unload their troops in the same general area. What the U.S. force finds is two enemy ships and one large barge with troops at just about the same spot the U.S. forces were to land. It was just 0200 hours when the U.S. destroyer cover spotted this activity on radar, no more distant than seven miles away. Being detected, the enemy attempts to flee the area as the destroyers open fire on them. In just 15 minutes a large column of fire is seen and the radar screen on the destroyer went blank.

After the troops are unloaded north of Ormac, the destroyers and LSMs turn to make a return trip to Leyte Gulf and it is then that two enemy planes appear. The enemy drops bombs on the U.S. ships and all miss their targets. A half dozen Marine fighters then come over the area and drive off the attackers. It is now 0700 hours on 12 December. One hour later destroyer *Caldwell* is attacked. Three enemy planes go at her. Two are driven off by the *Caldwell's* fire but the third comes at her low on the water on the port side and crashed into her at the same moment the *Caldwell* takes two bomb hits. She now has serious damage and is on fire. The fires are knocked down and *Caldwell* remains on line. The destroyer *Caldwell* has 33 killed and missing and some 40 wounded.

Now come more planes and destroyer *Coningham* is under attack as are other destroyers in the group. Several planes are shot down by the destroyers as the Marine fighters come over and take charge. The Marine CAP knocks down 11 enemy planes.

Two of our carriers are ordered to patrol the area skies at night and to hit the airfields under cover of darkness. This would help to prevent morning take offs against our forces. The record shows that at this time and in this engagement, we had sunk some 45 ships, including twelve tankers and a cruiser, the *Kashili*. We lost four ships and 39 were damaged. The suiciders as before, are becoming a problem. As Christmas, 1944 came, General MacArthur issues a statement. In short he says, "*The Leyte-Samar campaign is about over with some mopping-up to take care of.*" Enemy troops are still hiding out and at times are making contact with the U.S. Army's 77th Division. But the enemy is on his own. Word comes through that General Yamashita has written off Leyte as a loss and will make no more attempts to reinforce or resupply the area. Neither will he attempt to withdraw those troops which are now surrounded and fighting a lost cause. Some enemy troops try to leave Leyte under cover of darkness in an attempt to reach Cebu some 25 miles away and still in Japanese hands. The P.T. boats move into the area of Ormac and Leyte's west coast and continuously hit these small boats and barges carrying the enemy who is trying to get away from Leyte.

It will be the first week of May 1945 before Leyte is secure. The American casualties at Leyte are reported to be 3508 killed or missing, 12,076 injured. The Navy and Marine had additional casualties amounting to several hundred men in the Leyte area. One report of enemy dead was given as 80,557. Prisoners taken were some 830 men. The Japanese reported deaths of 48,790 with over 13,000 still hiding out in different areas. The Japanese are now on their way to failure in everything they try to do. General Yamashita would later say that the loss of Leyte made him realize that a victory for Japan was impossible and that they would no doubt fail in their attempt to defend Luzon.

15 December. On this day, the U.S. lands on Mindoro Island. Prior to the landing the usual bombardment took place. Hundreds of rockets were also fired onto the island. Mindoro is some 300 miles from Leyte to the northwest. As we approach the island, the Japanese will bring into play again their new defensive measure, the

kamikaze suicide planes. We see the carrier *Essex* take a hit from one of these suicide planes. We are after their air strip on Mindoro which will give us a take off point to hit Luzon when we invade it in less than a month. Mindoro is less than half the size of Mindanao, and less than a third the size of Luzon.

The landing on Mindoro was unopposed and the natives were friendly to the U.S. troops. They despise the Japanese for the mistreatment of them for some four years. General MacArthur will land on the beach with the troops and not one man will be lost. The island, like all others, received bombardment from the big guns prior to the landing. The bombardment was by the Seventh Fleet under the command of Rear Admiral Russel Berkey. The troops will cover and take the island completely and will not make contact with the enemy. Mindoro was bombed for ten days before the landings by heavy bombers of the 13th Army Air Force. The landing beach was small, only some 1400 yards long. The troops were taken ashore by Navy landing craft. Some fire was met by the troops as they landed but they went ashore without incident. One American P-38 fighter was shot down over Mindoro; the plane crashes and the pilot walks away unhurt.

With the latest action at Mindoro and the massive bombing of the airfields on Luzon, Task Force 38 groups pull back to refuel. At this time, the Army is in control at Leyte. Units of the fleet had been in Ulithi Lagoon on December 10 and from there headed to Luzon where the carrier planes hit many areas and wrecked many enemy planes on the airfields. Strikes were made on Manila also.

It is midday of December 16 and the fleet pulls back to refuel. The wind is making up, the sea getting rough. Refueling is difficult as more than once the fuel hoses between the ships are parted. The sea is too angry and so refueling is discontinued. The fleet moves to another area and again they try to refuel. Now the wind is much stronger and again refueling is put off and again the ships steam to another area. After three tries at refueling, it is cancelled indefinitely. There is no doubt in anyone's mind now: The fleet is in a typhoon which builds with every minute.

It is now midday of December 17. The ships toss and roll and pitch in the heavy seas. All ships try to tie down anything that might move or slide in the rolling and pitching ships. The carrier planes are lashed down with triple bindings. The torpedo bombers are brought below to the hanger deck and made secure. Bomb carts, tow tractors and fork lifts are all lashed down to the deck. Anything and everything is tied down and no one is allowed on deck unless on watch. Safety lines are strung all over the deck so that anyone on deck has something to hang on to. The wind is now 80 knots and increasing.

It is now early on December 18. Visibility is poor. Ships to the left and to the right cannot be seen. The wind is now over 100 miles per hour. All ships are directed by radar control.

The destroyers cannot keep a straight course. They rise, they fall, they slide sideways in the angry seas. The destroyers of the screen cannot be seen on radar as the huge waves all but bury them. Planes begin to break loose from their moorings and they slide off the flight decks. Planes on the hanger deck break loose also and they smash into other planes and into the bulkheads. Some fires erupt. The smaller carriers, *Cowpens, Montery, Cabot* and others are rolling 35 to 45 degrees in the swells. Sea water runs all over the ships' decks below the flight decks. Three destroyers roll over and are lost. Only 100 men are found the next day; the rest of the crews of the three destroyers, some 800 men, are lost. The destroyers lost were the *Hull, Spence* and *Monaghan.* Eighteen to twenty ships are damaged and need some sort of repairs. The fleet formation broke up and the ships are all off their stations. The typhoon of the past October 3 to 6 also did damage to some ships My carrier, the *Intrepid,* lost seven planes off the flight deck in that October typhoon, and sustained some structural damage also.

Intrepid missed this typhoon of December 18. She was headed for Hunters Point, California for repairs after having been hit by two bomb laden suicide planes. Said Admiral Halsey after the typhoon of December 18, "*I will never, repeat, never trust an aerologist again!*" Evidently the weather reports he received were anything but right. Because of this typhoon, the planned strikes for December 20 to 24 were cancelled. The damaged ships, now back in Ulithi, will be repaired there and the fleet will again move out in early January. Said former Admiral Samuel Morison in one of his (15) books (Vol XIII, p. 59), "*For some reason that goes deep into the soul of a sailor, he moans for shipmates lost through the dangers of the sea, even more than for those killed by the violence of the enemy.*"

During the Mindoro invasion one LCI is hit at the landing beach by a single Japanese attack plane. A big LST pulls along side and takes the troops off the burning LCI. Some troops are killed and others are burned real bad but they will recover. The doctors on the LST treat the burn victims. To land on Mindoro the transports and cover ships had to travel 600 miles past many Japanese bases including Mindanao, Cebu, and Palawan Islands, and it was at the final moments prior to landing that the Japanese Air Force attacked the landing force. One suicider came straight down at 90 degrees and hit the *Essex,* just as they hit the *Intrepid* last time. The sound of their engine, sometimes, is the first that anyone knows they are in a dive on the ships. Some guns cannot train at that angle (straight up) and the suiciders get through the flak and they make contact with the ships. Mindoro was secured on 28 January 1945. But even before Mindoro fell, Luzon would be invaded on 9 January.

After the typhoon of December 18, 1944 the fleet rested and resupplied and remained in Ulithi lagoon until the end of December. The typhoon's damage to the ships was repaired and all rested up for the next strike which was at Lingayen Gulf and another conflict with remnants of the broken Japanese Imperial Fleet. Admiral Halsey went on board some of the ships and awarded decorations to those men who performed gallantly during the typhoon of December 18. Also decorated were some of the various squadrons whose pilots had done such outstanding work in their contact with the many enemy planes and their pinpoint bomb runs on the Japanese bases. The *Intrepid*, although not there at the time, was also commended for her actions in the many strikes and engagements of the past 80-odd days.

Lingayen Gulf is the next area of action as General Mac Arthur plans a further thrust into the Philippines. His troops will land on the west side of Luzon at Lingayen Gulf. This landing will take place on 9 January. Admiral Kinkaid's force will cover General MacArthur's invasion force at Lingayen Gulf. They will leave the Leyte area, move north off Mindoro, pass by the Manila Bay area, and enter Lingayen Gulf. Some Army Air Force assistance is now available in the area and their planes help to cover the movement of the force.

The fleet at Ulithi prepares to move out. The carrier planes are to make repeated strikes on the airfields of Luzon and Formosa. They will enter deep into the China Sea and hit Japanese Naval forces reported in that area. Task Force 38 departs Ulithi lagoon on 30 December. Some refueling takes place while the fleet is underway headed to a position off Formosa. The weather again turned bad. Rain fell, the wind increased, and it became overcast. These weather conditions held for over three weeks.

The first carrier strikes were against the fields on Formosa and then strikes were made on Okinawa beginning late on 3 January and continuing on the 4th. Strikes were then made on Luzon, hitting Japanese installations there, having left the Formosa area during darkness and moving to an area off Luzon, arriving early next morning. The assault force in Lingayen Gulf was not hit by planes from Luzon or Formosa, but were attacked by kamikaze planes from an area further south. The reports state that some 25 ships were damaged and a number of them were sunk due to the kamikaze suicide attacks of 3 through 6 January. The assault force was made up of many troops from all of our secured areas in the southern end of the Philippines. The troops went ashore at Lingayen Gulf on 9 January and there was no opposition until they got further inland and then (like the Palau invasion) the Japanese concealed forces opened up on the U.S. troops.

The huge invasion armada (some 850 ships) sailed north through the South China Sea from more southern areas in the Philippines, close by enemy held islands and bases. And the trip north to Lingayen Gulf was hell; the invasion force was subjected to 14 Japanese air attacks in five days. Someone — he is not known — wrote the following day by day log of this struggle up the west coast of the Philippines:

> ***January 4.*** — Yank planes flew 230 sorties without loss. One Japanese plane was shot down by anti-aircraft fire. Enemy planes continually skirted the formation.

January 5. — Thirty planes shot down. Two enemy destroyers were located and one was crippled and the other left apparently sinking. Carrier planes flew 350 sorties.

January 6. — More Japanese planes destroyed. American planes sank small craft in Lingayen Harbor. Damaged three railroad bridges, and blew up a large ammo dump. Carrier planes flew 309 sorties.

January 7. — Three Japanese planes destroyed. Fifty tons of bombs were dropped on shore installations. More small craft were sunk and enemy troops were destroyed. Three tankers demolished, three fuel dumps and an ammo dump blown up. Fifteen buildings, including barracks, offices and storage structures were bombed and Japanese anti-aircraft positions blasted. Carrier planes flew 316 sorties.

January 8. — Carrier planes flew 315 sorties, but the results were not divulged.

From this log it is obvious that just approaching the invasion site was anything but calm. Before the actual assault on the Lingayen beaches, the battlewagons point their big guns and open fire on the shore. The bombardment is so fierce it vibrates the area. Nothing could survive this pounding. The LCIs with their rocket launchers send salvo after salvo into the island. The troops then go ashore as Admiral Kinkaid's Seventh Fleet continues shelling and General Walter Kruger's Sixth Army lands on the beach. The troops move inland and spread out looking for the enemy who isn't there. Further inland the troops make contact with the enemy in the rice fields. Resistance is light. The landings are made on the south and southeast coast of Lingayen Gulf.

The Japanese thought we would land at Manila because the carrier planes had made a bomb run there. This was a diversionary tactic. Instead we landed at the Gulf's south end much further north than Manila. The ships of the Seventh Fleet are scattered all over the area. There are about 850 ships in the force. This force includes two attack groups, a backup reinforcement group, a bombardment group, and fire support groups as well as our covering group which is the carriers. The battleships are the fire support groups and they go into action first to soften up the beach and the islands and to hit reinforced gun positions. This continues as the landing craft head to the beaches. The cover group, the carriers, send their planes over the islands and the landing areas and bomb and strafe gun emplacements and the areas where the enemy may be dug in or putting out fire power at or onto the troops.

The fleet carriers in the meantime, move north again to Formosa and the planes once again go up to hit the airfields of Formosa. These strikes are intended to reduce enemy flights and in most cases stop the enemy from taking off to hinder the landing operations. But some enemy planes got through at the start of the invasion and attacked the U.S. ships in the area. The escort carrier *Ommaney Bay* was sunk on 4 January and three mine sweepers working the area were also sunk. The LSTs which are nosed up to the beach and unloading are strafed by a few enemy planes, but not a single man is injured in these raids. The Japanese had shore guns trained on the beach areas and hit the ships as they came in and after they got to the beach. A strike force is sent to silence these guns and that is just what they do. They were quickly put out of action.

Halsey then moves into the China Sea. Task Force 38, under cover of darkness and under strict radio silence, moves west. The fleet is undetected at this time. Tankers now arrive and the fleet refuels quickly in an area known as Balintang Channel. It is now midday of 11 January, the ships are refueled and Halsey begins a drive of the fleet deep into the China Sea. He has a report that the two half-battleship half-carrier *Ise* and *Hyuga* are at Camranh Bay and he wants them and intends to finish them off. These two ships were part of Ozawa's Northern Force which Task Force 38 hit so successfully some two months earlier after the Leyte invasion.

The fleet now moves to a position to strike Camranh Bay. The two half-battleships are not there, but the strike force had many other targets, mostly shipping, in close to shore and in the outer waters of the Bay. Many strikes were made and several convoys were hit. Some of the ships sunk were nine tankers, which had a cruiser as escort and it , too, was sunk. When the strike was over, fifteen combat ships were sunk and so, too, were some 29 other ships, including tankers and merchant ships. With this strike terminated, the fleet puts about and heads back toward Formosa's northern coast. Other enemy ships were spotted and attacked and sunk, but the two elusive half-battleships were not seen as the second week of January ended.

After riding out another storm off Formosa, the Task Force's next target on 20 January will be Takao. When it was all over ten enemy ships had been sunk including five tankers. A few enemy planes got off the ground as the U.S. carrier planes got over the target area, but they were quickly shot down, some not getting completely airborne when they were hit. Several strikes were made on the area and one report states that 104 enemy planes were destroyed.

Task Force 38, which has Task Groups 1, 2 and 3 taking part, still had Admiral Sherman in command of T.G. 38.3 and Admiral Arthur Radford commanding T.G. 38.1. Task Group 38.2 (*Intrepid's* old unit) still has Admiral Gerald Bogan in command, whose flag was on *Intrepid* when she was put out of action on 25

November. The carriers are lined up off shore with Sherman's group to the north and Radford's group to the south and Bogan's group in the middle of a north-south line. The 21st day of January would be a bad day for Task Force 38. With so many ships having been damaged in the prior two months, new ships arrived to take their place in the fleet lineup. Two of these ships were the carrier *Ticonderoga* (CV-14) and the small carrier *San Jacinto* (CVL-30). It is a few minutes before 1200 hours on 21 January as a group of bogies appear on radar. At this moment the battleships *Washington* and *North Carolina* are fueling some of the destroyers with them. Other destroyers move in pairs to the north and to the south on picket duty and early warning of any enemy who may be headed to the task force. They are some 40 or 50 miles distant from other units of the force. they also would report back on our own planes returning from their strike at Formosa. There were two reasons for this set up. Our returning planes, as they showed up on that ship's radar, would then be known to be friendly. Another reason for the picket destroyers was to watch to see if any enemy planes followed the U.S. planes back to their carriers.

Two groups of bogies are now spotted on radar on their way to the task force; one group is given a position as being some ten miles ahead of our returning planes and the other some five miles to the north (bearing 330) of our returning aircraft from the Formosa strike. The battleship *Washington* reports these bogies on radar as friendly, but which bogies? The ones on course 330 headed for the task force or the ones on course 280 degrees some 64 miles away? Other ships are in disagreement with the *Washington's* report that the bogies on radar are friendly, but they do not state the range or course. Everyone now comes alive to deal with this situation. CAP and radar plot people begin to investigate the recent radar reports. Fighters are readied for take off to go to meet this blip which is on the screen. Is it friend or foe, or is it a group of enemy approaching at the same time our planes are headed back to the carriers? If the incoming force is friendly, we can recall the planes now leaving the decks. New checks are being made in radar plot to determine the incoming force as being friend or foe. The bogies are now at a 55 mile range. A message is sent to the flag ship (*Washington*) informing them that two groups are on radar. Our planes returning are a few miles astern of the other bogey and some to the south of it, but now both are on the same course toward the task force. In the center of this 55 mile range, are the destroyers *Cassin, Young* and *Calahan.* The *Washington* reports back that both bogies on screen are friendly and so no order is given to intercept. The bogies are now on course 110 at a speed of 180 knots. The blip on radar now splits into two groups. One group continues on the 110 course, the other moving to the south. The bogies should soon be over our picket destroyers. One group is still ahead of the other on radar. As our strike planes got over our picket destroyers they dropped down to a low altitude so that the destroyer would know that they were friendly. At this moment they disappeared from the task force radar screens and also at this moment the other group (enemy) swung back to the north. Carrier *Essex* had CAP over the carriers, and its fighter director at this time instructs the CAP to intercept planes approaching the task force.

One problem here was that our returning planes flew at 10,000 feet while the enemy flew at some 20,000 feet. Carrier *Langley* (CVL-27) spotted the first enemy plane coming in with the sun at his back and the *Langley* opened fire on it. CAP is not attacking these enemy planes. The first plane spotted by the *Langley* drops two 550 pound bombs. One goes into the sea beside the ship and the other bomb penetrates the *Langley's* flight deck exploding below on the hanger deck. The plane does not pull out of its dive and ploughs into the sea. The *Langley* now is afire and her flight deck is holed. The torn up flight deck area is 10 by 14 feet, a good size opening. The *Langley* was lucky as the bomb which hit her was a small one and hit in an area more or less free of stored ammo. Three men are killed and 11 injured on the *Langley* due to the exploding bomb.

A second plane now is headed for carrier *Ticonderoga*. He drops his 550 pound bomb on her flight deck; the bomb with its delayed fuse goes right through the timbers and explodes below. *Ticonderoga* is now afire also. The *Ticonderoga* has a half dozen planes fueled and armed on deck, ready for take off. They begin to blow up almost instantly after the bomb explodes. These ships are part of Admiral Sherman's Task Group 38.3. Another group of enemy planes of about a dozen head to Admiral Radford's group 38.1 to the south and to the east of Sherman's group, some 25 miles away. T.G. 38.1 is ready for them and the CAP takes care of them and no ships are hit. The planes that hit Radford's group came from Luzon fields. The group that hit Sherman's force seems to have taken off from Taito in south Formosa.

Radar now tracks another bogey group which also came from Formosa and it is now 75 miles and closing. With *Ticonderoga* afire and unable to take aboard or launch aircraft, her planes already in the air are directed out to attack the new incoming group of enemy planes. They are given instructions to fly at 20,000 feet while a second group off the light carrier *San Jacinto* is ordered to follow them at an 8,000 foot altitude. At a point 15 miles from the task force the *San Jacinto* planes make contact with the incoming enemy flight. Radar now shows the original bogey to be in two parts. The ones intercepted by *San Jacinto* planes has only four planes in the group and three are shot down. The planes are of the Oscar type. The fourth plane headed to the stricken *Ticonderoga* and crash dived the carrier, hitting the island structure. More fire for the crew of the *Ticonderoga* to deal with.

The *Ticonderoga* suffered. She had 143 killed and 202 wounded and lost 36 of her planes. Her skipper, Captain Kiefer, was one of the wounded and was badly burned from the fire which erupted when the suicider crashed into the island structure. Destroyer *Maddox*, on station, was also hit that afternoon and had many casualties — more than the carrier's. Admiral Sherman orders the *Ticonderoga* and the *Maddox* to Ulithi for repairs and orders an escort to go with them. The *Ticonderoga* will be out of action for three months. All this action took place in about 25 minutes, beginning just before noon of 21 January, 1945. 22 January is supposed to be the day an attack on Okinawa is to take place.

The *Langley* crew waste no time in repairing her damage after her fires are out. The hole in the flight deck is covered with heavy steel plates as we had done so many times on my ship, the *Intrepid.* Late on 21 January the fleet turns and heads north to Okinawa covering the distance during the night and early on the 22nd. Okinawa was becoming important. Landings on this island were planned for early April and we needed up-to-date, low level photos of the beaches, airfields, the inland terrain and military installations of all kinds. Strikes were launched before dawn on 22 January 1945 and six more during the day, devastating everything within range and in perfect weather (for a change). Then Task Force 38 was ordered back to Ulithi for repairs, replenishment of supplies, and a good rest, arriving there on 25 January for a two week layover. They would sortie from Ulithi on 10 February as Task Force 58, reflecting the relief of Admiral Halsey by Admiral Spruance. This was to be an eventful mission. The target was Tokyo.

The fleet passed Guam on the 12th and started gunnery practice but accidents marred the day; in flight, an errant rocket blew off the tail of a Hellcat and pieces of it tore up the wing of another fighter. Both pilots died in the crashes. The Task Force refueled on the next two days and late on 14 February the long run-in to Tokyo began as the weather turned colder in the more northern latitudes. The speed northward was increased on the 15th and finally early on 16 February, a little over 100 miles from Tokyo, the force was in position to launch the first fighter strike of the war against the Japanese home islands — against Tokyo itself. The weather was not good for the first strike; clouds to 15,000 feet and it was cold. The first group of 20 Hellcat pilots couldn't even find the Japanese coast, but the second group of 15 did and had an unopposed field day bombing and strafing airfields near Tokyo. Parked planes were flamed easily and at least a half dozen others were shot from the sky within minutes. They had achieved total surprise. As more strikes were launched the air opposition over the Tokyo area increased and the fighting soon became a wild melee and before noon some 20 enemy planes had been flamed in the air and destroyed. This was getting close to another turkey shoot; the Japanese pilots, although flying some excellent aircraft, seemed totally combat-inexperienced. Some could fly well and even do some fancy acrobatics, but many didn't even fire their guns when they had a chance for a hit on the attacking Hellcats. They flew like the idea of deadly combat panicked them. The air defenses over and around Tokyo and the remains of her once great fleet were so weak that the enemy couldn't even mount a serious attack on the carrier force, barely 70 miles away. Some Helldiver dive bombers were sent over the city in the afternoon, looking for targets that had been reported by pilots of the morning's fighter sweep.

On 16 February 1945, we returned to our most bitter defeat of the early war in the Pacific — back in May of 1942 following a four-month siege by the Japanese — when our Army on Corregidore Island, at the entrance to Manila Bay, finally succumbed. Their treatment of the American military prisoners and of civilians was at best inhumane. Army units with the Navy assisting now invade the island. Some men come onto the island by parachute in the hilly areas as others hit the beach. The island received a massive bombardment from the big guns of the cruisers and battleships and it was bombed and strafed by the carrier planes. Gun emplacements, buildings, ammo dumps, oil storage areas — all were hit. Army Liberator bombers also made runs over the island hitting key defensive areas. The paratroopers had some casualties and some overshot their landing point and came down in the water of the bay. Navy boats went out to pick them up. It took about two weeks to secure the island, far shorter than the four months it took the enemy to capture it.

There were problems with hidden gun emplacements and stubborn Japanese defenders. Some 6,000 Japanese troops were on the island and many were inside their fortified caves and bunkers. Just about every Japanese soldier had to be killed since they would not surrender — as usual. Some were captured and at one count there were 4,215 Japanese bodies which were bulldozed into a mass grave. The U.S. casualties were 136 killed, 8 missing and 531 injured. The Japanese had us two to one in strength and they fought to the last man. They tried several banzai charges in an effort to stop or slow the U.S. advance in its drive across the island. Corregidore, in the center of the approaches to Manila, had to be retaken or the Japanese could have wreaked havoc on our ships and men entering the area. Japanese guns on the island could hit Manila and other areas around it. After some 21 months the stars and stripes now fly again over Corregidore.

Dawn of 17 February saw a continuation of the carrier strikes against Tokyo and its environs. The torpedo squadrons got the "go" signal this day. The U.S. Navy now owned the skies over the Japanese homeland it seemed, since enemy fighters, although airborne in large numbers this day, did not appear to have the stomach to attack the incoming Hellcats. So the F6 pilots took the fight to them and plane after plane with the big red ball on their wings crashed into the frozen, snow-covered countryside. The first strike of the day netted dozens of confirmed air kills and many more destroyed on the ground in strafing runs. Task Force Helldivers and Avengers were then launched to hit aircraft factories north of Tokyo. Enemy fighters were up in force but the F6 pilots did not allow a single one to penetrate into the bomber formations. More Zeke, Tony, George and Tojo fighters were shot down as cloudy weather and rain squalls moved into the area and the Task Force withdrew southward. Not one plane of the carrier fleet was lost to enemy action this day. The two days of attacks, the first by a carrier fleet on Japan's home islands during the war, had been a resounding success: At least 340 enemy planes shot down and perhaps close to 200 destroyed on the ground! The entire fleet lost 60 planes and nearly as many airmen. To add more injury to injury, massive B-29 strikes from the Marianas hit Japan at the same time that Task Force 58 was sweeping clean her skies of the defending air force. Both the Japanese military and the Japanese citizen now knew that the days of the Empire were numbered.

These strikes at the Japanese homeland were in part a cover operation for one of the bloodiest island assaults of the war: The Marines' invasion of Iwo Jima. It was believed that the enemy would oppose these landings with all the air power they could muster, bringing in suicide planes from Tokyo, 640 miles away, if necessary. As it turned out, they were almost powerless to do anything; although there would be losses to kamikaze attacks, the skies over Iwo Jima belonged to the United States Navy.

The fleet ran south on 18 February to Chichi Jima in the Bonin Islands, 500 miles away, refueled, and went into action again. Iwo Jima in the Volcano Islands, another 140 miles further south of Chichi, was on the Marines' hit list for the 19th; they would need all the helping fire power they could get. Iwo Jima is a volcanic island with black ash as its soil, and with an extinct volcano, Mt. Suribachi, at its south end. The pear-shaped island is about five miles long and two and a half miles wide at its widest (south) end. The Japanese have dug into this volcano and have gun emplacements in caves trained on the beach. These guns would take many lives on the beach and a strong first effort is made to silence them. For three days the big guns pounded the island fortress and thousands of rockets were sent against the volcano. The 14 inch guns of the battleship *New York* pounded the island and the hill also for 48 hours prior to the landings.

The beach is only two miles wide and becomes overcrowded as the 4th Marine Division on the right, and the 5th Marine Division on the left assault the island at 0900 hours. As the Marines stormed ashore at Iwo, the carriers of T.F. 58 hit the airfields and shipping on and around Chichi Jima; there was another carrier force of 16 CVE (escort) carriers supporting the landings on Iwo and T.F. 58 would not be needed there — yet. You couldn't tell that to the Marines on Iwo this day; they were being slaughtered on the eastern beaches of this miserable little island and wondered why the big carrier fleet was still over a hundred miles away and not helping them. They perhaps didn't know it, but Iwo had been pounded so mercilessly preceding the assault, that there were simply no visible targets above ground to be hit. The Japanese defenders, 20,000 strong and fanatic, were all underground in caves and tunnels. Never-the-less, the T.F. 58 carriers sent off fighter strikes on the 20th to bomb and rocket suspected enemy troop concentrations on Iwo. More strikes from T.F. 58 were launched on the 21st as the attacking Marines were now getting a toe-hold in the black volcanic ash which was the soil of Iwo Jima. The napalm-bombing fighters had to streak in low to guarantee a hit on their targets — cave and tunnel entrances and pill boxes — and were returning to their carriers with their bellies scorched black by their own fire bombs.

As mentioned above, the 4th and 5th Marine Divisions were selected to invade Iwo Jima with two regiments of the 3rd Division in reserve. 800 ships of the U.S. fleet are there and 30,000 to 40,000 troops are offshore and ready. As noted above, the big guns of the battleships and cruisers blast and bombard the island, prior to the landing, for 72 hours. The ships circle the island and all guns fire from all directions. The LCIs pour rockets on to the island also. Lt. General Holland M. Smith is commanding the Marines as he has in other invasions. The signal is given to go in on the morning of 19 February 1945. Landing craft used were LCIs, LSM's and LSTs. Troops were also landed from small boats from the transports offshore to the small eastern beach. These smaller boats, after dropping the troops on shore, picked up the wounded and took them back to the transports which had a staff of doctors on board. Waiting for the men to get a foothold and move off the beach, the AKAs and APAs wait to come in and unload supplies, food, ammo, etc. As far as you can see there are ships everywhere in all directions. The Fifth Marine Division begins their move through the volcanic ash to begin the assault on Mt. Suribachi.

As noted earlier the beach is littered with wrecked landing craft hit by the Japanese shore guns. The beach was exposed to a withering enemy fire and now the men are coming ashore in amtracs. These amtracs could maneuver on water or land and they protected the men inside them until they got to the beach; in many instances the men used them on and off the beach like small tanks. They were lightly armored and protected the men from small arms fire. They also had 50 caliber machine guns mounted on them. The amtracs made the going a little easier as they moved through the deep volcanic ash which was everywhere.

The Japanese were well dug in and hidden and opened fire on the advancing Marines. The carrier planes were called in on the second day and again on the third day to strafe and bomb the hidden enemy. Iwo Jima would be listed as the bloodiest battle to this date in the Pacific for the Marines. The Marines head toward the south airfield and it is soon taken over by them. The carrier planes make pinpoint hits on Mt. Suribachi trying to knock out the gun emplacements on the sides of the hill. By night fall and during the dark hours more and more men and supplies are brought ashore. By 1900 hours there would be 30,000 men ashore, all in a foxhole of some kind. When the sun went down on this first day, 566 Marines would be dead and another 1,755 wounded. The hospital ship surgical rooms, designed to handle no more than 20 serious cases a day, saw three times that number on this 19 February.

The enemy was underground in caves and tunnels, in pill boxes and bunkers. The two regiments of the 3rd Marine Division, the offshore reserve, were called in on D+2 and placed in the center of the line between the 4th and 5th Divisions. It will take three weeks or more to blast, burn and hand grenade the enemy inside. Some caves will be set with charges under cover of darkness and then blown up to seal them. Many caves were sealed but openings between the rocks gave them peep holes to fire from if they survived the blast. 5,931 Marines died on Iwo Jima in the 35 days of combat, 19 February to 26 March; 17,272 more Marines were wounded. 881 Navy men were killed on board ship, in the air and ashore and another 1,917 wounded. On 16 March mop-up operations begin and will take some ten days to complete. Of the 20,000 Japanese defenders on the island (conservative estimate) less than 900 were taken prisoner; the rest died. My brother, Corp. Carmen F. Zollo, will be on Iwo as a forward observer with the big 155 mm guns of the Fourth Marine Division Amphibious Unit. He will radio back to his lines the positions and type of fire power the Japanese had behind the lines. He will do well and will not be injured.

Iwo was important. The island lies about half way between the B-29 bases in the Marianas and the Japanese homeland. The big bombers either had to swing wide around Iwo or accept being spotted on their missions and reported to Tokyo. Iwo would also provide a base for long range fighters that could accompany the B-29s to their targets. And finally, this half-way point would provide an emergency landing field for damaged bombers returning from their mission. In this latter capacity, Iwo, by war's end in August would have taken in 2,400 crippled B-29s and their 26,000 crewmen. Many of these — the statistics say 50% — would have been lost at sea had Iwo Jima still been in Japanese hands.

But not everything was going the American way. As the Marines on shore were suffering great casualties, some enemy planes began arriving — perhaps from Chichi, some certainly on one-way trips from Japan itself. Two kamikazes and several bombs went straight into the flight deck of the old *Saratoga*, CV-3, setting her afire and she withdrew from battle. Then a smaller escort carrier, the *Bismarck Sea*, CVE-95, was sunk by kamikazes with the loss of over 300 men. From sundown on several bogeys attempted to penetrate the T.F. 58 formation but all were driven off by the fleet carrier's 5 inch guns, if they made it past the battleship and destroyer screen. The carrier task force sent a fighter sweep to both Iwo and Chichi Jima on the 22nd and then headed north to punish Japan again.

The weather worsened and the Task Force slowed down in the high seas. Finally, on 25 February at about 200 miles from Tokyo, the carriers launched their fighters. About an hour later at 0830 the sweep of 48 Hellcats arrived over Tokyo with some three dozen enemy planes in the air waiting for them. These defending pilots had not learned much from the previous strike; they performed some fancy flying but were not aggressive, and they died, one by one. The weather got worse and worse as the "ceiling" lowered to nearly wave-top level. All further strikes were cancelled and, now low on bombs and ammunition, the Task Groups turned south away from Tokyo and set a course for "the barn", Ulithi, arriving there on 1 March 1945.

Intrepid was at Hunters Point until about the middle of February, 1945. Repairs were completed early that month and on 11 February she went to sea for post-repair trials, gunnery practice, air operations and calibration of equipment, returning to Alameda on the 13th. On 15 February, Captain Giles E. Short relieved Captain Joseph F. Bolger as *Intrepid's* Commanding Officer. The following day Air Group 10, under command of Commander John F. Nyland, reported aboard for duty and plane loading began immediately. Air Group 10's equipment consisted of 37 Hellcat fighters (VF), 30 fighter/bombers (VBF), 15 Helldiver bombers and 15 Avenger torpedo bombers. At 0920 hours on 20 February the big carrier headed for Pearl Harbor, on her way to war again, arriving at Pearl on the morning of 24 February. On 3 March 1945 she departed Hawaii as part of Task Group 12.2 in company with the carriers *Bataan* and *Franklin* and eight destroyers; destination: Ulithi.

Two days before *Intrepid* arrived at Ulithi, on 11 March at about 2000 hours, three kamikazes screamed in low over the hundreds of ships anchored in Ulithi's lagoon. For unknown reasons, one crashed into the sea and a second crashed into one of the surrounding islands. But the third, a twin engine Frances, singled out the *Randolph*, CV-15, and crashed with a devastating explosion into her after flight deck. Twenty five men died and 106 were wounded in the blast and fire that followed. In an hour the fires were out and it was all over. Ulithi was not that far removed from combat. The source of these suicide planes could only be guessed at, then; it later was discovered that about 25 of these suiciders had taken off from Minami Shima in the Volcano Islands, near Iwo Jima, some 1,000 miles from Ulithi. Only the three had successfully navigated the distance to Ulithi, the remainder getting lost, crashing, and their crews probably dying at sea. The *Randolph* would require extensive repairs. Test alerts were staged periodically for the next few days to keep everyone on their toes; no one wanted another disaster like the attack on the *Randolph*.

Captain Joe Bolger is relieved of Command on the Intrepid 15 February, 1945. (National Archives Photo)

Captain Giles Short relieves Captain Joe Bolger on 15 February, 1945. (National Archives Photo)

The days of the *Intrepid's* long westward passage to Ulithi were filled with constant sea routine duties: gunnery practice, flight operations and a thousand duties necessary to keep a complex ship like a carrier in top operating condition. *Intrepid* arrived at Ulithi on 13 March, the task group was dissolved, and *Intrepid* reported for duty to Admiral Marc Mitscher, Commander Carrier Task Force 58, in *Bunker Hill*, for duty. The following day *Intrepid* was assigned to Task Group 58.4 and on this day, 14 March 1945 departed Ulithi for battle. With *Intrepid* in this Group were the fast carrier *Yorktown* (CV-10) and the light carriers *Langley* (CVL-27) and *Independence* (CVL-22). Nine more fast carriers and four more light carriers made up the rest of Carrier Task Force 58, a grand total of 17 carriers to take the fight to the enemy. Admiral Spruance's flag is on the cruiser *Indianapolis*.

Intrepid is Back
Invasion of Okinawa

Early on the morning of 14 March 1945 the *Intrepid,* one of two large carriers, plus two light carriers of Task Group 58.4 under Admiral Radford, weighed its 15 ton anchor and headed northwest. The mission of this enormous carrier force (11 heavy and 6 light carriers, to be joined later with some two dozen escort carriers) was to support the assault on Okinawa, scheduled for 1 April 1945, two weeks away. Okinawa was the most southern island of the Ryuku Island chain extending from the Japanese home islands. This was the beginning of the invasion of Japan itself, and it would not be a walkover; the enemy was expected to defend his home ground with a fanaticism not yet seen in this war. *Intrepid's* Air Group 10 practiced their landings during the next two days, and everyone was tense as the Task Force approached the Ryukus on the 17th. Radar was picking up enemy planes far from the force, but none attacked; in the early morning hours of the 18th an F6F night fighter was vectored to one bogey that got too close, and shot it down. This day would be a long one for *Intrepid* and the rest of the carriers. The Japanese would contest the Okinawa landings with all the air power they had left, and that meant kamikaze attacks on the fleet. The suicide attacks by these green and orange-painted "zootsuiters" (so named after the wild colors worn by some American teenagers at the time) were feared most by everyone from seaman to admiral; if the surrounding screen allowed them to penetrate to the carriers, and if the carriers could not destroy them or damage them severely on their suicide run, someone was going to get hit. This attack technique was not like dropping bombs or torpedoes which would often miss the target or could be avoided; the kamikaze was a human-guided missile that had a high probability of a hit if its pilot were still alive and didn't panic at the last second.

It is now dawn of 18 March, and the planes leave the deck to hit Kyushu, Japan. The fleet is about 100 miles offshore as the first strike takes off. This raid of 18 March will see over fifty strikes on enemy targets and the enemy will attack the fleet also, many times during the day, with a mixture of different types of planes. The attacking Japanese planes still come in low just above the water as they did so many times before; but they also come in very high as they began to do back in November when my ship, the *Intrepid*, was hit. CAP was constantly on the alert for the high flyers.

There was no set pattern to the enemy attacks on the fleet. One flight could have bombers in it with fighters and some could have kamikaze units with them. It was a busy day, dawn to dusk. Radar and CIC plots were kept on their toes all day, plotting and directing the planes. When an enemy is approaching the fleet, sometimes the CAP and the radar people have only minutes to break up the attack before the enemy reaches the ships. Not always is this possible and so a plane or two

The Yorktown under attack off Kyushu, Japan on 18 March 1945 as seen from the Intrepid. (National Archives Photo)

Jap suicide plane is hit by Intrepid guns and crashes beside the ship. (National Archives Photo)

A Japanese plane has just been shot down by Intrepid's guns. (National Archives Photo)

Intrepid turns hard to port to shoot down a Japanese attacker. (National Archives Photo)

Intrepid is attacked by a suicide plane, as seen from OL-49. Other carriers and battleships try to repel the attackers. (National Archives Photo)

Enemy plane coming straight in on Intrepid, as two are shot down. (National Archives Photo)

Suicide plane can be seen just before he made a dive on the Intrepid. (National Archives Photo)

Enemy plane shot down by Intrepid on starboard rear. (National Archives Photo)

Intrepid hit and afire as seen by CVL-22 Independence. (National Archives Photo)

As she burns, the Intrepid's guns hit a plane who explodes to her right. A destroyer stands by the injured Intrepid. (National Archives Photo)

Intrepid afire on the starboard side hanger deck. (National Archives Photo)

The Intrepid burns, with all planes spotted forward. Several crash barriers are up to stop planes - ours or the Japanese who may land on the deck. (National Archives Photo)

will get in, and if it's a kamikaze, he will crash the ships; and those enemy planes down low on the water are many times unseen by radar. One plane that got through dropped a bomb on carrier *Yorktown* ; it was a Judy and the ship had five killed and 26 more wounded. It was 1309 hours.

Earlier my ship, *Intrepid,* has a Betty bomber make a run on her. From the action report for this day:

At 0805 Task Group guns began firing at a twin-engine bomber, presumably a "Betty", and set it afire. The plane attempted to crash into the ship; however, the guns of this ship hit the plane with particularly heavy fire from 40mm and 20mm guns causing the plane to go out of control and crash into the water about twenty-five to fifty feet opposite the bridge. It exploded with shrapnel effect throwing its burning gasoline tank into the ship along with other parts of the plane. At 0806, Repair 7 reported fire on the forward end of the hanger deck. The temporary alcohol stowage, movie screen, fuel oil hose and other miscellaneous matter were ablaze. ....While the ship was under attack by the plane mentioned above, a 5" projectile from another ship of the Task Group burst in the vicinity of number 8, 5" single mount causing damage to mount, minor shrapnel damage to surrounding hull structure and casualties to personnel. Simultaneously a 20mm shell from another ship damaged 40mm Quad #12.

There was one death and 44 other casualties in all these actions. No suicider got to the ships this day, but tomorrow is another day.

On 18 March 1945 the landing on Panay Island, between Mindoro to the north and Negros to the south, is pulled off by the U.S. Army. The Philippino natives and the guerrillas are friendly to the invading U.S. troops, and assist them in many ways in hitting the Japanese on the island. In two and one half days the island is secure. Our P.T. boats were active here at Panay. They spotted and took care of the Japanese resistance which was close to the shore at different places around the island and at Mindoro. The guerrillas helped also to get equipment and supplies ashore. The guerrillas were armed by the U.S. troops and they assisted them before and after the landing on Panay.

Our strikes found some good targets during the day. Ships were hit and one of the ships was the 72,000 ton super battleship *Yamato.* This ship was hit twice before and on one of these times, it was hit by *Intrepid* planes in the Sibuyan Sea when *Intrepid's* planes hit *Yamato's* sister ship, the *Musashi* and sank it on October 25, 1944. The *Yamato* now again has *Intrepid* planes over her near the coast of Honshu, Japan, and again the planes score as the *Yamato* is hit. The airfields were hit hard and many planes wrecked on the ground at Kyushu.

It is now early on the 19th of March and *Franklin* is launching planes and a lone enemy plane is over the ship as *Franklin's* planes are taking off. Two bombs are dropped on *Franklin,* one goes through to the hanger deck, the other explodes on the flight deck in the middle of *Franklin's* planes, manned and armed and warming up before take off. Other planes below on the hanger deck are also ready to come up topside and take off on a strike. Planes below and topside blow up and fire is everywhere. Two of the three elevators are put out of commission. Within minutes, a second undetected enemy plane drops a bomb on carrier *Wasp,* not far from *Franklin* . *Wasp* had already secured from flight operations, her planes were already in the air. The fires which erupted on *Wasp* were quickly put out but she had over 100 killed and some 260 wounded. *Franklin,* meanwhile, is a raging inferno. Other ships come along side and assist the *Franklin* in fighting her fires. The *Franklin* had 724 of her crew killed and missing with 265 others wounded. The fires, after a time, are knocked down and the cruiser *Pittsburgh* puts a line on *Franklin* and tows her some distance away. A little later Franklin, under her own power, turns for home, and she arrives eventually at *New York* almost 12,000 miles away from where she was hit. She was some sight to see as the tugs took over and brought her to a dock, as a dry dock was prepared for her. Admiral Davison was *Franklin's* group Commander and he sent orders to *Franklin's* skipper, Captain Leslie Gehres, telling him to prepare to abandon the ship; but Captain Gehres stayed with her and they fought the fires and the carrier was saved.

As we have done so many times before, as a strike or raid or bomb run is completed for the day, the ships retire overnight to a safer area and those needing fuel are fueled. Sometimes we would pull back as many as 250 to 300 miles at flank speed to a predesignated meeting with the oilers. The oilers always had escort carriers to protect them and these carriers each could have as many as 40 planes on board. Many times, some of those planes were launched and were taken on board the fast carriers to replace some of their missing aircraft. No longer do the ships of the task force have to go back to Ulithi for ammo and provisions as the supply ships now travel with the tankers. No longer do we have to run to the store, you might say; the store comes to us.

On 21 March the Task Force is still off Kyushu, Japan and they expect the Japanese to try to hit them and that's what the Japanese try to do. They now use mostly twin engined Betty or Frances bombers with some Zeros for cover. General Quarters sounds, bogies 85 miles and closing, coming from the northwest. *Intrepid* fighters and others go to intercept this group of enemy planes which is now said to be a large group. Now more planes are launched to assist those already in the air and headed to the enemy group. The U.S. planes fly high at perhaps 18,000 feet and as they near their target, come down to the enemy's altitude and attack. The enemy force is made up of bombers with fighter escorts. The U.S. and Japanese planes meet

60 miles from the Task Force. One hell of a dog fight takes place. Then our pilots see something new. The Betty bombers have the new and strange-looking "Baka bomb" slung under their fuselage. The bomb is let go from the Bettys with a human inside guiding it to the U.S. ship targets. Our planes photograph these strange looking bombs under the Betty bombers and they got some photos of the Baka bomb as it was released from the bombers.

With so many U.S. planes now at the contact area, the Japanese planes are shot down in a really furious battle. Some of the Betty bombers dropped their piloted Baka bombs into the sea and were going to attempt to retire. It is reported that this action lasted fifteen minutes and in that time every one of the Betty bombers was knocked down and all but two of their fighter escorts also were shot down. After the plane photos are examined and studied, and from later enemy reports, we find that this group of Betty bombers numbered eighteen and they had thirty fighters with them.

On 22 March the Task Force refueled and rearmed and moved south into position to hit Okinawa, leaving the Honshu, Kyushu area. Target date for "Operation Iceberg" — the invasion of Okinawa — is 1 April 1945. The bombing of the Kirama Retto islands will be first and then the southeast coast of Okinawa. Landings take place on Easter Sunday, April 1, not on the southeast end which the carrier planes hit, but on the western coast near the two airfields. Kirama Retto will be taken first. These are a small group of islands a short distance to the west of Okinawa. These islands will be invaded on 26 March. It is expected that the Japanese will mount many large scale attacks on the U.S Task Force. The kamikaze will be well represented in these attacks. 182,000 troops wait on the transports and all their supplies and needs are on supply ships. Some 433 assault ships will take the troops ashore. The Okinawa invasion will be the largest operation of the Pacific War — and the last.

As mentioned above Kirama Retto was invaded on 26 March after the islands were given a good going over by the big guns and by the carrier plane bombing and strafing runs. Kirama Retto had a suicide boat base and this was eliminated by the striking planes. To neutralize expected attacks from the Kyushu area, B-29s hit that area on 27 and 31 March. On 29 March, Task force 58 left the Okinawa area and went back to Kyushu area to hit its airfields again. Resistance was light as the troops landed and occupied the Kirama Retto group. Then the bombing and strafing runs began anew on Okinawa. The big guns of the battleships and cruisers opened up in a tremendous assault. Then, early on April 1, the troops went ashore, and resistance was light in the beginning. Casualties were as expected but this would change in the next few days. During these first few days, the enemy attacks did come in the form of kamikaze raids; these early attacks were repelled and no ships were hit.

The 6th Marine Division lands on the shore opposite the Yontan airfield and to their right (south) lands the 1st Marine Division. To the right of the 1st lands the

7th and then the 96th Army Divisions. This puts the 96th Division opposite the Katena airfield. The 6th Division will turn left and head north through the Yontan airfield area, taking the airfield en route. The 1st Division will push straight ahead to cut the island in half. The 7th will also move inland for a ways to the outskirts of Katina airfield, then turn right, taking the airfield on the way. The 96th will turn right at the shore and head for the Machinato airfield. They will be at the narrowest area of Okinawa and opposite Nakagusukie Bay. Meanwhile the 6th division heads to the northern tip and on to the Motobu Peninsula. They will encounter fierce resistance there.

The 1st Division is now some three miles inland. The 7th Division, after penetrating deep inland, turns south toward the area where the 96th Division is operating near Machinato airfield. These two divisions meet up and spread out in a line and continue their drive south. They soon are stopped by the Japanese who have drawn a defense line across this area from Naha, the capitol city, to Shuri and to Yonabaru (an airfield) at Naha. The Asa River flows eastward at this point and the Japanese figured this would stop the advancing Army troops. The Shuri and Yonabaru areas are rugged and this too, the Japanese figured, would stop the Army drive southward. The fighting became more and more intense. Casualties were mounting on both sides.

By this time the 1st and 6th Divisions had headed south in a sweep from the Yontan and Katina airfields area and just about meet up with the two Army divisions. The 6th Marine Division moves to the west of the 96th Division, and the 1st Marine division moves into the Machinato airfield area which is just north of the Asa River. At this time it is 1 May and a month of bitter fighting sees the Allied forces far from conquering Okinawa. By the 9th of May the 6th Marine Division moves to the west and is now also at the Asa river on the far right (west). Between May 3rd and 5th the Japanese make a strong stand and counterattack in the Asa River area. The Japanese used sea and air forces to repel the line of U.S. troops stretched out on the north side of the river. The enemy manages to get some of their forces behind the American lines. The infiltrated enemy forces are wiped out and in the early hours of the next morning, 10 May, the Marines cross the Asa River about a mile from the capitol of Naha. By the next day the Marines are on the outskirts of the city.

The Japanese resist and drive the Marines back about 200 yards, but the Marines rally and push through the Japanese assault line. During all this movement and action the Marines and the Army troops had to overcome pillboxes of concrete, log bunkers, artillery installations, some rough terrain and caves, not to mention the Asa River. The honey combed caves in the Kushi-Yake mountain areas were a big problem. The caves were deep and had a series of tunnels connecting them underground. The only effective weapons on these were flame throwers and hand grenades. On the approaches to Naha is a hill named "Sugar Loaf Hill." It more or less protects the city. It was here that the Marines took severe and heavy casualties. There were eleven assaults on this hill before the U.S. troops could go over the top and down the other side to Naha.

It had been raining for eleven days and the land was a quagmire. This slowed the advance and made for a supply problem to the troops. The Japanese moved to the east side of Sheri, buying time. On May 29th the 1st Marine Division captured Sheri Castle, outflanking the Japanese, and soon the entire Sheri area was cleared of enemy troops. Meanwhile, the 6th Marine Division, in an amphibious move, land behind the Japanese lines and fight their way to the airfield. It is now 6 June. Next day they overran the field and went southeast and met up with the 1st Marine Division. In a last ditch stand the Japanese gathered their forces and moved to a high hilly area on Yaeju Dake. This area was 600 feet high and some 3,000 yards long. The fighting here was fierce. One by one the enemy positions on the ridge were taken. The 8th Regimental Combat Team of the 2nd Marine Division was brought into action and helped in the final drive of the enemy to the sea. By June 21, resistance had ended.

The invasion of Okinawa took some one million soldiers, sailors, Marines and air men to secure the island. 5,990 Americans — Army, Navy and Marine — and killed in this vicious battle, and another 17,000 wounded. 165 more are listed as

(l. to r.) Lt. (jg) Leroy Bartels, J. S. Austen PHOM3/C, E. J. Watowski PHOM3/C and E. F. Dalzal CPHOM study a map while on the Intrepid 2 April, 1945. (National Archives Photo)

"missing" and presumed dead. But we now have a large base on Japanese home soil, less than 1,000 nautical miles from Tokyo. It would be the jump-off point for the final invasion.

On 6 April, a massive fleet of kamikaze planes and others appeared over the Task Force. In the group were fighters, and Betty and Frances bombers. Reports later will state that these planes came from the Kanoya area at Kyushu. The same area was raided by the Task Force earlier and again on 29 March, and was also hit by the B-29s on two other days. Over 100 enemy planes skirted the Task Force and headed right to Okinawa to hit what ever they could including the ships close in to shore and the landing and assault forces on the beach. Supply ships were hit as were some destroyers who were in close shelling the inner beach areas. The Task Force had destroyers on picket duty as the outer screen, and they served as early warning and were the first to open up on the attacking force. With all the planes of the U.S. carriers in the air at the time, they were still out-numbered by the enemy attack force. Some of these attacking planes went to the destroyer screen, some directly to the landing site and the rest tangled with our CAP and other planes which took off from the carriers. This latest force of attacking planes came from a new group based at Kanoya, Japan and consisted of some 700 planes, some of them bombers and some fighters and 355 of them were kamikaze suicide planes.

Sixty of these planes were attacked by the Fleet's Combat Air Patrol as they neared the Task Force's position. As noted above, 100 headed to the invasion area on Okinawa. Estimates made later would say that 182 planes came over the Task Force. CAP shot down 60 of them; 24 crashed in the water after being hit. The ships' guns shot down 35 more. Reports state that a total of 249 enemy planes were destroyed in this first attack against our forces at Okinawa. Three destroyers were hit as was one LST and two ammunition ships and they were all sunk. Ten other ships were hit and damaged. Now a new threat is in the works as the Japanese prepare the super battleship *Yamato* with nine other ships and send her to Okinawa to do battle. And she has only a one-way ticket — her fuel is sufficient for only getting her there. It seems the *Yamato* is about to become a kamikaze herself.

On 6 April, 1945 the 72,000 ton battleship *Yamato* leaves her berth and heads to Okinawa. It is about one hour before sunset as two American submarines in the Bungo Strait area detect the monster battleship *Yamato*, with one cruiser and eight destroyers coming out of the Inland Sea. A message is sent to Admiral Nimitz at Pearl Harbor informing him of the ship sighting and her course. Admiral Spruance is alerted at the same time. Spruance orders Admiral Deyo at Okinawa, who is the big gun commander, to ready battleships and cruisers to go and intercept the *Yamato* and her escorts. The *Yamato* at over 72,000 tons, compared to our battleship Missouri's 45,000 tons, was something to see. She had nine 18.1 inch guns, the largest of any ship in the world. Her sides had 16.1 inches of armor plate. She carried over 3,000 men in her crew and she could do 27 knots. She also had 100 anti-aircraft guns on her. Her mission was to engage and draw away the Task Force at Okinawa. With this done, a massive air attack was planned to hit the troops, ships and anything else the U.S. had at Okinawa.

One mission was to put the airfields, now in U.S. hands, out of commission to deprive the U.S. of airfields that could be used to strike back at Japan. The two U.S. subs that spotted this *Yamato* group of ships were the *Threadfin* and the *Hackleback.* The *Yamato* is commanded by Vice Admiral Seiichi Ito. The *Yamato* had left Tokuyama Bay some two and a half hours prior to her sighting by the U.S. subs. My carrier, *Intrepid* will have a hand in the encounter with the *Yamato* just as she did with *Yamato's* sister ship, the *Musashi*. With the big guns of Admiral Deyo getting underway, so too does Admiral Mitscher turn the task Force around and heads north hoping to strike the *Yamato* force early next day. At dawn, search planes go up to look for the *Yamato* force. One scout plane from *Essex* spots the force and reports back their speed and heading. Two PBMs go up and they will shadow the Japanese ships for a time, reporting back any changes in direction. They stayed aloft for several hours. The cruiser with *Yamato* is now known to be the *Yahagi*, and she is moving at 22 knots.

We now decide that this is another bait and draw maneuver of the Japanese, such as the Ozawa northern force at Engaño when we vacated the Leyte area to pursue his force. Now the *Yamato* force is the bait to draw the fleet away from Okinawa so that a massive aircraft attack can be made on our forces at Okinawa. Now, the *Yamato* turns west away from Okinawa to the east. The spotter planes report that the *Yamato* and her escorts are moving in a circular pattern. The *Yahagi* leads, the *Yamato* in the center, with the eight destroyers circling them. The Task Forces launches the first deck load of planes. Eight carriers from Task Groups 58.1 and 58.2 send up 386 planes. A little later, another 109 planes take off from carriers of Task Group 58.4 leaving many planes behind on CAP duty.

The first group to go into *Yamato's* force was from T.G. 58.1, which has carriers *Hornet*, *Bennington*, *Belleau Wood*, and *San Jacinto*. Some clouds help to protect the planes as does the overcast condition which prevails. The first ship hit is the *Yamato*. Two 1,000 pound bombs fall on her and at the same moment, a torpedo strikes her port side. Within minutes, the cruiser *Yahagi* is also hit with a torpedo and takes a direct bomb hit and she is dead in the water. The destroyers scatter and one is torpedoed and sinks in minutes. All this took place in a matter of minutes, as the first strike turns to leave and the second strike dives into the fray. The planes of wave one were of T.G. 58.1. Now, planes of T.G. 58.3 pick up the strike. These planes of T.G. 58.3 are from carriers *Essex*, *Bataan*, and *Bunker Hill*. They drop a half dozen torpedoes at the *Yamato* and five hit her; another torpedo bomber comes along and puts another torpedo in her starboard side. Three more torpedoes hit her

Japanese suicide plane hits the Battleship Missouri, as seen from the Intrepid. (National Archives Photo)

Intrepid has port list as she burns after attack. (National Archives Photo)

port side. She now is listing and is taking on water. Other planes are now going after the *Yahagi*. She takes seven hits from the torpedo planes and twelve hits by the dive bombers. The *Yahagi* sinks in less than thirty minutes. The attack began at 1241 hours and it is now 1345 hours. At 1400 hours, the planes of T.G. 58.4 come in and by 1420 hours the *Yamato* is sunk. Four destroyers were sunk also, and two others were badly damaged and retired from the area. We lost ten aircraft and twelve men in the *Yamato* strikes.

While this action took place on this 6 April, 200 miles from Okinawa, a kamikaze force of 41 planes escorted by fighters was hitting the carriers. With the kamikazes were some Army planes which went after the shipping at Okinawa, both at the beach and offshore. CAP had a busy time protecting their carriers. Carrier *Hancock* was hit as was a destroyer and some of the battleships. The battleships never got to do their thing, as the carrier planes finished off the *Yamato* force.

Things were quiet for a few days and the Task Group refueled and rearmed on the 9th and 10th. On 11 April kamikaze assault #2 came at the Task Force and at Okinawa. This force had 185 kamikaze planes escorted by 150 fighters and it also had 45 torpedo bombers with it; a massive strike force which went first for the carriers. Many were intercepted before they got to the ships, but some got in and the *Essex, Enterprise,* and the battleship *Missouri* were hit along with three destroyers. The following day (12 April) destroyer *Abel* was kamikazied and minutes later was hit by a Baka (human) bomb. The destroyer went down quickly. When this attack ended, fourteen U.S. ships had been damaged and the *Abel* sunk. 298 enemy planes were shot down in these two days of attacks.

Early on the 13th of April, bad news arrived. President Roosevelt had died on April 12, at Warm Springs, Georgia. All ships' ensigns were lowered to half mast. Vice President Harry Truman is now our new President.

It is now 14 April and Betty bombers carrying Baka bombs come at the Task Force. They are quickly brought down by the CAP. A strike is set up and the carrier planes hit the southern end of Kyushu. Some 80 planes are destroyed with no U.S. losses. It is now 16 April and kamikaze strike #3 comes at the carriers, 165 planes in this group. This time many of them go for the outer ring of destroyers. But *Intrepid* gets hit again. A suicide plane plows into the *Intrepid's* flight deck. Damage is severe and again *Intrepid* will go stateside for major repairs. The following excerpts are taken from the action report dated 23 April, one week after the kamikaze hit:

General Quarters sounded at 1327; Condition Zebra was set throughout the ship at 1332. ....The ship was intermittently under attack from the time General Quarters sounded until 1336 at which time a distinct shock was felt throughout the ship. A Japanese suicide plane had crashed with its bomb into the port side of the Flight Deck, between frame 147 - 152. The bomb exploded on the Hangar Deck at

Yamato Force is Destroyed

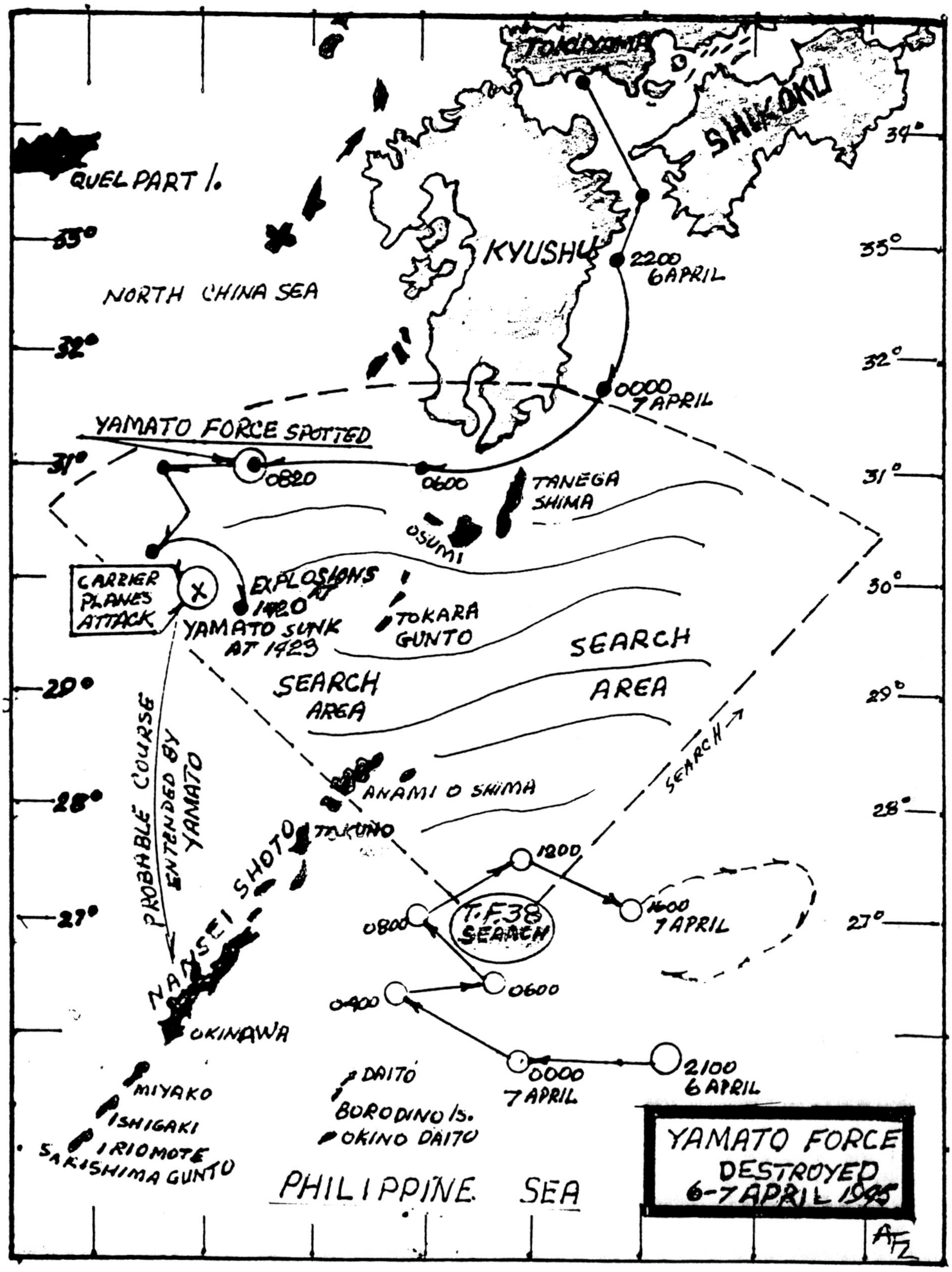

frame 145 causing a conflagration which extended from frame 100 aft to frame 160. the Hangar Deck was immediately filled with smoke. .....Personnel from Repairs 2, 3, and 7 and subsidiary units of Repair 1 laid out hose and began fighting the fire which was largely confined to Bays 3 and 4. .....As a result of the inability of Repair parties to secure the sprinkling systems immediately (electrical control systems had been destroyed), a great amount of water collected on the Hangar Deck. A temporary list, by which this water and burning gasoline was cascaded over the port side, was obtained by starboard turns. .....At 1426 all fire was reported as being extinguished. .....By 1640 the hole in the Flight Deck had been temporarily repaired and the ship was then ready to land aircraft.

There was extensive damage to the flight deck, the gallery deck and the hangar deck, and to aircraft. The bomb carried by the suicider was believed to be a large 1100 pounder. The hole in the flight deck was 20 by 15 feet and the deck was lifted up on either side of the hole. No. 3 elevator was damaged and out of commission. Some arresting gear was extensively damaged. On the Gallery Deck various workshop and squadron armory and ready rooms were severely damaged. On the Hangar Deck where the bomb exploded, damage was extensive, demolishing completely some frames. Gasoline servicing systems, water sprinkling systems, fuel oil systems, ventilation ducts, steam and air pressure supply systems — all severely damaged. Most of the electrical control systems for the elevators, some gun mounts, sprinkler systems and lighting were destroyed. Radar and radio remote control systems were damaged. 40 planes were destroyed before the fires could be knocked down. Eight men were killed in the bomb explosion, and 21 others injured. The gallant *Intrepid* had been back to the war less than a month.

Enemy plane is shot down, and its bomb broke lose on contact with the water and exploded. (National Archives Photo)

A near miss at Intrepid's starboard side sends flame and hot metal onto her decks. (National Archives Photo)

Intrepid is in a bad way as her own planes explode aft. (National Archives Photo)

Hanger deck fire still smoking as the clean up continues. (National Archives Photo)

The smoke begins to clear after experiencing a near smoke black out after the crash of a enemy plane. (National Archives Photo)

The Intrepid is afire after the attack. All ships take evasive action. Planes are returning from their bomb run over Japan. Photo taken by CV-20 Bennington's photo plane. (National Archives Photo)

94
85

Most of the rubble is cleared away, as the hanger deck continues to smolder . (National Archives Photo)

Two of the 45 planes that were damaged on the hanger deck. (National Archives Photo)

Elevator pit is flooded from all the water put on the fires. (National Archives Photo)

Eight men of the Intrepid are put to rest at sea after the 16 April, 1945 attack. (National Archives Photo)

One more to bury at sea as the sun begins to set. May they rest in peace. (National Archives Photo)

Intrepid Heads Home Again

Intrepid could no longer stay on line with her extensive damage and she turned south toward Ulithi, arriving there on 20 April 1945. Repairs were begun immediately with the intention of returning the big carrier to action. Ammunition, fuel and stores were shipped aboard in preparation for completion of the repairs. But on 28 April, after a great deal or repair work had been accomplished, #2 elevator was discovered to be out of alignment. On 30 April the repair crews decided that the *Intrepid's* extensive damage was too much for their facilities to handle, and she was ordered home. 48 planes were catapulted off to go to other carriers, and her ammunition was unloaded. On 3 May passengers and pilots of Air Group 45 from the *San Jacinto* reported aboard and at 0800 on 4 May *Intrepid* left Ulithi for Pearl Harbor. She passed Eniwetok on 7 May and arrived at Ford Island, Pearl Harbor at 1740 hours on 11 May. After refueling and unloading superfluous ammunition and bombs, *Intrepid* steamed out of the channel at Pearl at 1700 hours on 14 May and headed northeast. On 19 May she entered the channel at San Francisco and by 1300 hours she was moored at Alameda Naval Air Station.

Starting on 7 April 1945 the beaches on the southwest end of the island of Mindanao were bombed continuously for some nine days running. On the 17th, Mindanao is subjected to its final bombardment as the troops of the Army's 24th Division land at Parang and Malabang beaches on Mindanao's west coast. Resistance was very light and the troops moved inland. They reached the capitol of Catalato and in less than three days had it secured. The guns of Rear Admiral Russell Berkey's unit of the 7th Fleet did the bombardment of this area. This area is 200 miles south of Leyte. Also assisting in the bombing were many bombers of the 13th Air Force. We reported that the resistance was light, and it was, but the Japanese had guns trained on the beaches and on landing craft as they approached. No landing craft were hit. The Army divisions were under command of Lt. Gen. Robert J. Eichelberger and led by Maj. Gen. James A. Dee. The fighting at different areas of Mindanao drags on into May, but mopping up is complete by the end of the month.

Days go by with no action against the fleet by the kamikazes until 27 April. 115 kamikazes come at the ships. This is strike #4 and nine more ships are kamikazied. On 28 April, the hospital ship *Comfort* is hit by a kamikaze. On 29 April, two destroyers are hit; they were the *Hazelwood* and *Haggard* and both are badly damaged. The score to this point was seven U.S. ships sunk and 56 damaged by these kamikaze raids. Other ships had some minor damages also attributed to the suicide attacks. As for the enemy planes lost, the number by the end of April was 820. These losses are attack losses and do not include the miscellaneous planes shot down or destroyed between the kamikaze raids, nor does it include those planes hit on the ground when the U.S. planes hit the airfields at Kyushu, Okinawa and other areas.

Intrepid's scoreboard as of 1 May, 1945. (National Archives Photo)

Kamikaze raid #5 comes on 3 May. This time, two destroyers are sunk and two LSMs and fourteen other ships are hit and damaged. On 7 May word comes that Germany has surrendered; finally, some good news. May 11 brought kamikaze attack #6, 150 of them. Carrier *Bunker Hill* is hit twice and she retires to the States for repairs. She was Flag for Admiral Mitscher and he moves to *Enterprise*. Three days later, on 14 April, the *Enterprise* is hit and Admiral Mitscher is forced to move again, this time to the *Randolph*. Some ships, during April and May, manage to get to Ulithi for repairs and some R&R. Every third or fourth day some of the ships refueled. Two more kamikaze attacks, #7 and #8, were warded off in May, but one ship was hit which later sank. In mid May the carriers made another strike on the Japanese home islands.

At this time carriers *Intrepid, Bunker Hill* and *Enterprise* were all headed stateside for damage repairs. Admiral Spruance in overall command of Task Force 58 is now relieved by Admiral Halsey and Task Force 58 is redesignated Task Force 38. In the first week of June 1945 the Task Force suffers through another typhoon and some ships are damaged. 13 June finds the fleet at Ulithi where damages are repaired and the men get some R&R at Mog Mog Island. Some two weeks later, all is set to go back to the action and the fleet sails and heads west to the Japanese home waters where more strikes are made. The score for the two months: 2,336 enemy planes destroyed by the carriers (including those on the ground); U.S. planes lost —

557. U.S. ships sunk: 30; U.S. ships damaged — 368 (includes all sizes smaller than the destroyers). Japanese kamikaze planes lost in ten raids on U.S. ships: 1,465. Adding others shot down between raids brings this total to 1,700. The *Yamato* and the *Yahagi* and four enemy destroyers were sunk. Operation Iceberg, the assault on Okinawa, is about complete. Okinawa was secured on 22 June, after seven weeks of bitter fighting.

One More Strike for *Intrepid* and War's End

On 20 May 1945 *Intrepid* crossed the Bay to Hunters Point and on 27 May was dry-docked for repairs. The ship's crew was given leave in shifts, but leaves would not be long ones this time. All repairs were completed to satisfaction in less than a month, *Intrepid* was floated, sailed to Alameda to pick up Air Group 10, refueled and rearmed, and on 29 June 1945 steamed out of San Francisco Bay for Pearl Harbor. She arrived at Pearl on 5 July and for the next three weeks sortied to and from Pearl on training exercises. By 30 July she was ready to go back to war and with Captain Giles Short in command and in the company of two destroyers, *Intrepid* headed for Eniwetok, via Wake Island. It was over 2,000 nautical miles westward to Wake, and another 530 south southwest to Eniwetok. *Intrepid* arrived due south of Wake late on 5 August and Air Group 10 flew a day of aerial strikes at Wake on the 6th. Four fighter sweeps and two bomber strikes — a total of 164 sorties (plus 18 CAP and ASP launches) — were made against water distillation and power plants and gun positions and a bridge on the island. They dropped 42 tons of bombs on the two small islands (Wake and Peale) and fired down 490 5" rockets. Two bombers were lost at sea in operational accidents but all their crewmen were rescued. (Wake was a favorite target — a kind of <u>must</u> — for any combat ship that happened to be near it for any reason; Wake was reduced to rubble time and time again during the war. The little islands had no strategic importance but the periodic strikes gave our gunners and pilots hands-on practice). The next day *Intrepid* arrived at Eniwetok.

The first atomic bomb had been dropped on Hiroshima (6 August) as Intrepid was raiding Wake Island, and the second would be dropped on Nagasaki (9 August) while she was in port at Eniwetok. Rumors of surrender and peace began to fly back and forth across the Pacific and across the U.S. and *Intrepid* not-so-patiently remained at Eniwetok until everything got sorted out.

While the *Intrepid* was being repaired and during her return across the Pacific, Halsey's Third Fleet was running wild over Japan. Their targets are anything

Intrepid Throws The Last Punch - Wake!

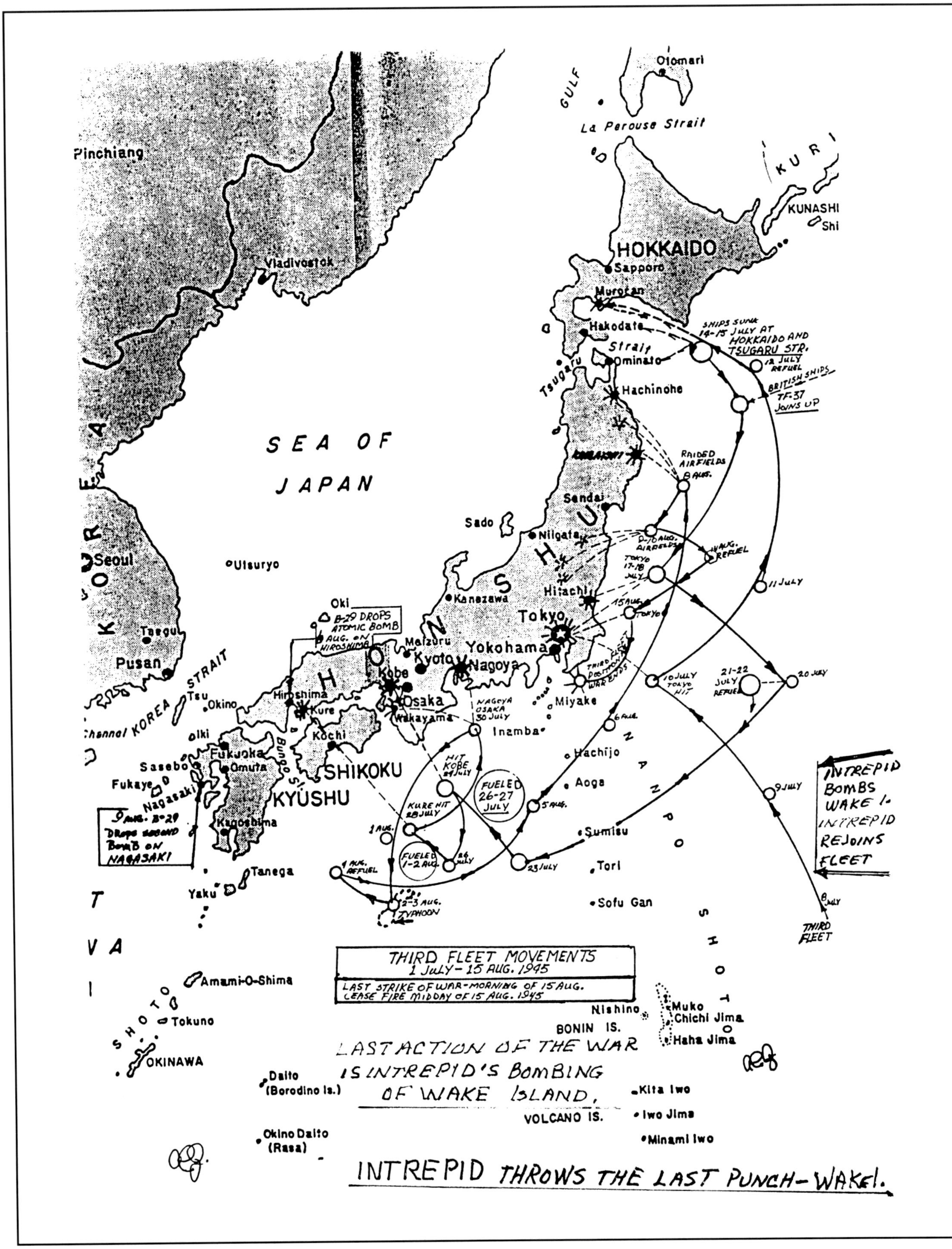

military such as oil tanks, ammo dumps, piers and docks in commercial areas, shipyards, factories, air fields and naval bases; anything and everything is struck. The areas most hit were in and around Kure, Yokosuka, Kyushu Island and Kobe to name but a few. The rivers and harbors were covered and anything seen was hit or, if floating, sunk.

In the strikes of late July the battleship *Nagato* is sunk at Yokosuka, the battleships *Haruna* and *Hyugo* are sunk at Kure, and in a second strike on this same day, also at Kure, the battleship *Ise* is sunk. On the 28th of July at Kure we sink the light cruisers *Kitagami* and *Oyodo* and the destroyer *Nashi.* A few days before, this time with Army B-29s, the carrier planes blast the Japanese Naval Arsenal and the naval base at Kure with the big bombs, the one-tonners. The Japanese do not have too much going for them at this time. On the 28th of July two heavy cruisers, the *Aoba* and *Tone* were sunk at Kure. The Japanese lost what was left of their carrier fleet also in this rampage. The large carriers *Amasgi* and *Katsuragi* were sunk at Kure. These played no part in the final months of the war; the Japanese had few planes to put on their decks. On the 24th of July the small carrier *Kaiyo* was sunk in the Japan Sea. These U.S. carrier strikes continued almost unopposed into the first two weeks of August. We lost a few planes to anti-aircraft fire and to the few fighter planes the enemy was able to put up, but resistance was half-hearted.

The final fighter and bomber strikes by Task Force 38's planes off Japan, against airfields and a huge electronics manufacturing and assembly plant near Tokyo were launched between 0400 and 0600 on 15 August; for the 150 mile flight the TBM torpedo bombers were loaded with big 2,000 pound bombs. Then, at about 0630, Task Force 38 received a radio message directly from Admiral Nimitz' headquarters which was relayed immediately to the planes en route for the day's strike: The Japanese had accepted the terms of the Potsdam Declaration. Cease hostilities, return to the ship, dump your ordnance in the ocean. Japan had surrendered.

The Japanese had accepted, verbally, the Allied terms of surrender on 14 August. The official instrument of surrender is presented to the Japanese officials by General MacArthur at Manila on 19 August. A typhoon delayed the Third Fleet's approach to Japan for the next few days, but units of the Fleet moved into Sagami Bay southwest of Yokosuka on the 27th of August. On the 29th other Third Fleet Units with Admiral Halsey enter Tokyo Bay, and Admiral Nimitz arrived from his headquarters on Guam and went on board his flag ship, the battleship *South Dakota.* On 30 August 10,000 Marines were landed at the Yokosuka Naval Base. On the 31st U.S. troops begin going ashore at Tokyo, and on 1 September more Marines occupied the Tateyama Naval Base, across the bay from Yokosuka. Everything is now in place and the meeting of the Japanese and American military commanders is now at hand.

MacArthur signs for the military.

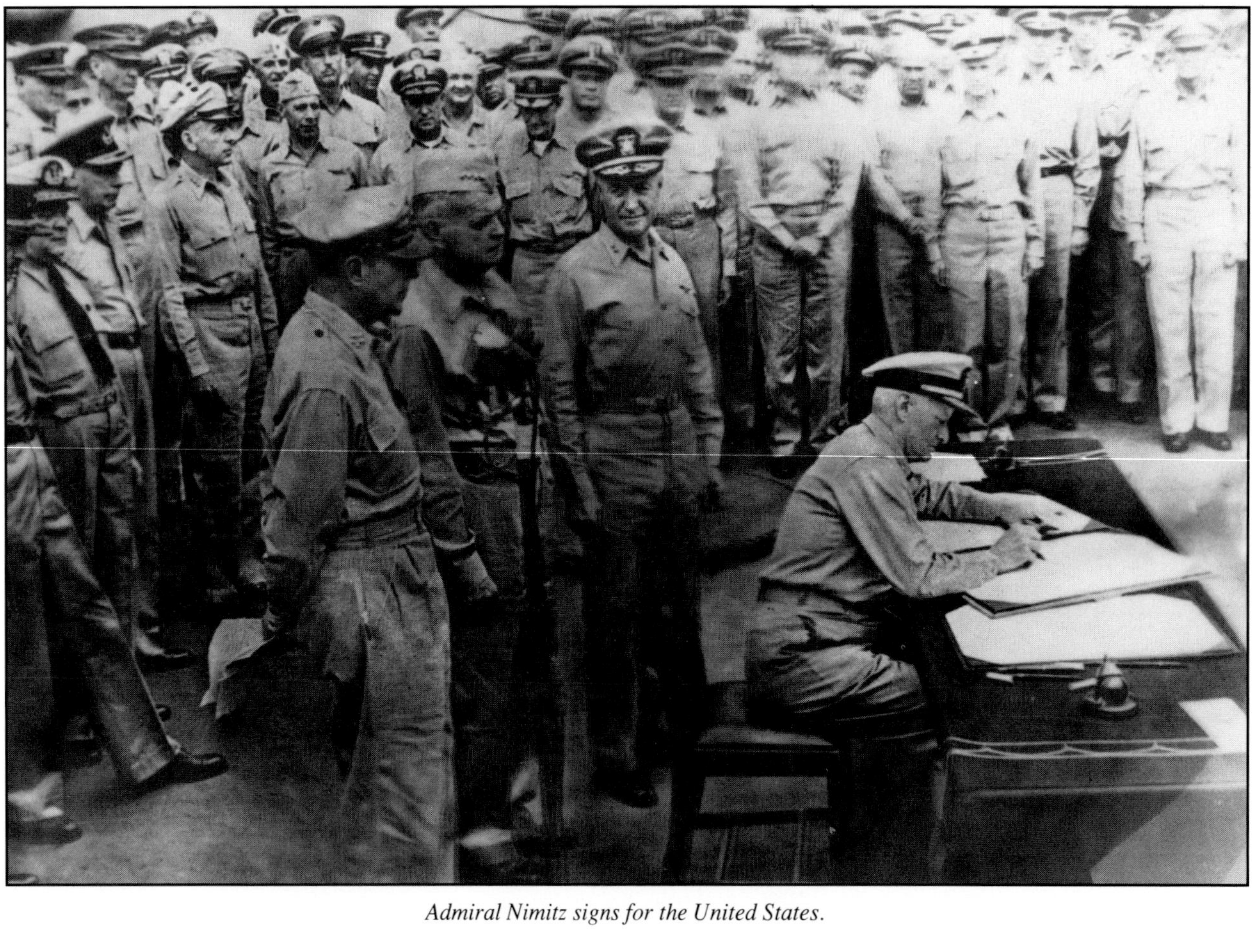

Admiral Nimitz signs for the United States.

The signing of the surrender by the Japanese.

The Japanese delegation awaits their orders.

Intrepid planes fly over Shanghai on show of strength patrol in September of 1945. Intrepid planes now have a white stripe on the tail, instead of the white cross used during the war. (National Archives Photo)

At 0908 hours on 2 September 1945 on board the U.S. battleship *Missouri* (representing President Truman's home state) the surrender of Japan to the Allies is formally signed. General Douglas MacArthur signs as the Supreme Commander of the Allied Powers in the Pacific, and Fleet Admiral Chester Nimitz signs as the representative of the United States. Even before the signing of the surrender document Japanese commanders all over the Pacific laid down their arms and surrendered to American forces. The islands which we repeatedly bombarded and bombed and raided but never invaded, began to heed the surrender terms. The Japanese Army commanders gave up Truk, Wake, the Palaus, Ponape, Jalluit,

Maleolap, Wotje, Rota, Pagen and all others. Truk alone surrendered 130,000 troops. Marcus and Mile in the Marshalls also surrendered. Though defeated, Japan still had four million men under arms and better trained than it had four or more years ago. Japan surrendered unconditionally and without even token resistance.

Never before and perhaps never again will there be such a force and such determination as there was here in the Pacific in those times of the early 40s on up to the Fall of 1945. Had the atomic bomb not been dropped, had the Japanese refused to surrender, then the world would have seen the mightiest naval force ever begin the largest sea borne invasion ever, and the defeat by force of the Japanese Empire. The Japanese chose to surrender.

Commodore Perry first opened up Japan to the outside world in the mid 19th century and now, almost 100 years later, the navy again opened Japan to the outside world by its defeat in the Pacific by American forces. World War II began in Europe on September 1st 1939 and ended in Japan on September 2nd 1945. World War II resulted in the greatest cost in human life, military and civilian, ever recorded. A conservative estimate places the number at 55 million. Some nations in Europe lost one tenth of their population.

Intrepid passed her second birthday, 16 August 1945 at anchor in Eniwetok lagoon. She had the dubious distinction of being the "most hit" aircraft carrier in the Navy. At the end of hostilities, now only days away, *Intrepid's* Air Groups will have shot down 160 enemy planes and destroyed 86 more in ground attacks. They will have sunk 11 enemy ships (plus two probables) and damaged at least 40 more.

The following is taken directly from Action Reports of the *Intrepid* during these final days of the war:

From 7 August to 21 August the Intrepid was at anchor in Eniwetok and the crew felt that she had been forgotten in the confused rumors of peace. When the Japanese accepted the Potsdam Declaration of terms for peace our future became anyone's guess. However, on 21 August, Commander Third Fleet ordered the formation of Task Unit 30.3.9 consisting of the Intrepid, the Antietam, and the Cabot, and 7 destroyers, under the Command of Captain Giles E. Short, United States Navy.

This unit sortied Eniwetok with orders to join the Third Fleet east of Japan. While enroute on 23 August the Antietam discovered structural damage and was detached with orders to sail to Guam and investigate the damage. Two days later the remaining ships of Task Unit 30.3.9 arrived and reported to Commander Task Force 38 and Commander Task Group 38.4 for duty. Three days later on 28 August the Intrepid received orders to proceed to Okinawa, to report to Commander Seventh Fleet and Commander Task Force 72 for duty. Task Unit 30.3.2 was formed consisting of the Intrepid, the Cabot, and the Blake, and the Chauncy, under command of Captain Giles E. Short, United States Navy, for this trip. The Task Unit arrived at Okinawa on 30 August 1945 and was dissolved. The ship reported to Commander Task Force 72, under the command of Rear Admiral A.C. Davis, United States Navy, for duty.

Intrepid anchored in Buckner Bay on 31 August and received fuel oil and aviation gasoline. She was then ordered to sortie from Okinawa and headed into the Yellow Sea off the coast of China, accompanied by *Antietam* (CV-36), *Cabot* (CVL-28) and five destroyers. Between 2 and 11 September this force operated off the China coast, sending flights westward to many cities in a "show of strength." It was important that everyone know that America and its Allies had complete control of the seas and the skies. The Japanese had been forbidden to operate in the Yellow Sea — ships or aircraft — and no opposition was encountered. In over 700 sorties from *Intrepid* there were six operational accidents during this ten day period in which six aircraft and two men were lost. On 13 September *Intrepid* returned to Okinawa.

Intrepid was resupplied and refueled for the next two weeks at Okinawa. On 27 September 1945 she was ordered out of Buckner Bay to the Yellow Sea to again "show strength" over China. She would be on station until 8 October. During this period flights were again made over many Chinese cities including over Shanghai, Tientsin and Peking. On 30 September and 1 October *Intrepid* provided air cover for the 3rd Amphibious Corps' landing in the Taku area; on 7 and 8 October air cover and reconnaissance were flown for U.S. Marines advancing from Tientsin to Peking. In the eight flying days during this period, *Intrepid* launched 679 sorties from her flight deck.

At sunset on 8 October *Intrepid* was released from her Task Force (72) and was ordered to Saipan. On 14 October she was anchored in Tanapag Harbor, Saipan. She was in port for only three hours, long enough to disembark Air Group 10 and take on Air Group 14, and then was en route to Guam. She arrived at Apra Harbor, Guam at 0800 hours the next day (15 October) this time for a five day stay for refueling and replenishment. On 21 October *Intrepid* up anchored for a return to Tokyo. During her four day trip northward she was launching and recovering her planes each day, keeping her aircrews in top shape. On 25 October *Intrepid* anchored in Tokyo Bay.

Intrepid steamed out of Tokyo Bay on 5 November in company with the *Lexington* (CV-16), two light cruisers and 11 destroyers for a series of intense training exercises in simulated torpedo attacks, gunnery practice, and bombing practice, all about 100 miles off the east coast of Japan. This force returned to Tokyo Bay on 9 November and conducted port routine until 23 November when the force sortied again for another round of intensive training offshore. No one on board was ever allowed to get 'rusty' in performing his duties. It was a lot of work. By 27 November *Intrepid* was back at her berth in Tokyo Bay.

By 1 December *Intrepid* had her orders to go home. She took on board almost 2,000 Army troops for transportation back to the States on this day. She refueled on the next day and at 0730 hours on 3 December, accompanied by the *Lexington, Intrepid* was on her way to San Francisco. It was a non-stop passage, due west (Tokyo and San Francisco are nearly at the same latitude), passing 1,000 miles north of Hawaii, and arriving at Alameda Naval Air Station on 15 December 1945. It was a happy crew that departed *Intrepid* for leave. On 18 December *Intrepid* embarked 1100 Army, Navy and Marine personnel on an overnight trip to Long Beach (San Pedro), California.

Decommissioned
Back to Duty with the Atlantic Fleet

Intrepid spent the next six weeks at Long Beach and then steamed north to San Francisco Bay on 4 February 1946, skippered by her new captain, R.E. Blick, USN. Blick was in command until 11 April when Captain H.G. Sanchez, USN took command as *Intrepid's* seventh skipper. On 15 August her status became "in commission, in reserve" and on 22 March 1947 *Intrepid* was decommissioned (*laid up* is the preferred Navy term) and placed in the Pacific Reserve Fleet. She had been on active duty for over three and one half years, and much of that time had been pure hell.

The big carrier languished in this state for the next five years, until the United States, as a member of the United Nations, became deeply involved in the Korean War. On 9 February 1952, Captain B.B.C. Lovett, USN in command, *Intrepid* was recommissioned just long enough to sail her to Norfolk, Virginia where she was decommissioned again on 9 April and work began in her conversion to a modern attack carrier. On 1 October 1952 her new designation became CVA-11 and, modernization completed, she was recommissioned, but in reserve, on 18 June 1954. On this date, Captain W.T. Easton, USN became her new skipper. On 15 October *Intrepid* went to full commission (for the third time in her history) as an operational

The Intrepid in moth balls prior to being recommissioned in 1952. Most of the other ships seen with her were eventually scrapped.

unit of the Atlantic Fleet. *Intrepid* was back. Just two days before joining the Atlantic Fleet, *Intrepid* had the distinction of being the first carrier in history to launch aircraft using American-manufactured steam catapults.

This modernization program (SCB-27C) involved some major changes in the carrier. Her full load displacement was increased from an original 36,380 tons to 41,944 tons. 18 feet had been added to her overall length, her beam on deck was increased from 147 feet, six inches to 192 feet, and her loaded draft was now 31 feet (from 28 feet, 7 inches). Besides her two new steam catapults, *Intrepid* now was fitted with two deck-edge elevators (from an original one) and one center-line elevator (from an original two). And with the introduction of jet-powered aircraft, her aircraft types changed along with a lot of new services that these types required. Her flight deck was reinforced to support operations of the heavier jets. Nearly 2,000 tons of jet fuel must now be on board, along with 60,000 more gallons of gasoline for the new and more powerful (and fuel hungry) generation of piston engines. The carrier now went to sea with a total of about 70 aircraft, mixed piston engine and jet powered, compared to her old maximum of about 100. And with all these changes *Intrepid's* crew compliment was reduced to 135 officers and 2,040 men (from 360 and 3,088 in 1945).

The rest of 1954 and the first months of 1955 saw *Intrepid* in shakedown cruises operating out of Guantanamo Bay, Cuba. With this cruise completed CVA-11 was assigned to duty with the Sixth Fleet in the Mediterranean and departed for those waters on 28 May 1955 with Air Group 4 on board. Her new skipper was Captain G.L. Kohr, USN. Kohr would be succeeded by Captain P.P. Blackburn, USN on 15 August 1955. This would be the first of two deployments to the

Mediterranean with the Sixth Fleet. On her second trip in March 1956 she had Air Group 8 on board. These Mediterranean excursions were essentially "show of force" cruises in that volatile part of Europe. *Intrepid* returned to Norfolk early in September of 1956 and by the end of the month was at the New York Navy Yard for modernization and overhaul (SCB-125). As of 21 September *Intrepid's* new commander was Captain F. Massey, USN.

This program of modernization was completed on 2 May 1957. Another eight feet was added to her overall length but the most obvious external change was the addition of a new angled flight deck, the hallmark of all new and modern carriers in the 1950s. The angled deck was used primarily for landings, while the center-line deck now functioned exclusively in catapult launching of aircraft. Shakedown and training was conducted out of Guantanamo Bay during the Summer of 1957 and then the modernized CVA-11 was ordered to European waters in September to take part in NATO Operation Strikeback, the largest peacetime naval operation ever conducted. *Intrepid* operated out of Norfolk as her home base and in December conducted an exercise named Operation Crosswind which examined the effects of various wind speeds and directions on aircraft launches. She was able to show that even the heavy jets could be safely launched at any wind angle, even downwind if necessary.

Through 1961 *Intrepid* operated in the Mediterranean, the Carribbean, and along the U.S. east coast as CVA-11. In the Spring of 1960 *Intrepid* was chosen to conduct launch trials of the experimental Phantom II fighter. In the last week of April, on the first launch of the fighter at full take-off power, the airplane overran the catapult and damaged it. The engines of this new fighter were enormously powerful and it was obvious that the aircraft really didn't need all the power available for take-off. Take-off thrust was reduced and all subsequent launches went smoothly. On 8 December of this year *Intrepid* was reclassified as an anti-submarine warfare support carrier, redesignated CVS-11, and entered the Norfolk Navy Yard on 10

An aerial view of Intrepid underway sporting her new angled flight deck and her new larger forward elevator.

The Intrepid undergoes conversion again so that she can carry bigger and heavier planes in preparation for the Korean War.

On Intrepid's deck, a North American AS-2, the first nuclear bomber is being loaded with bombs 17 November. 1954.

Intrepid at the Artic Circle

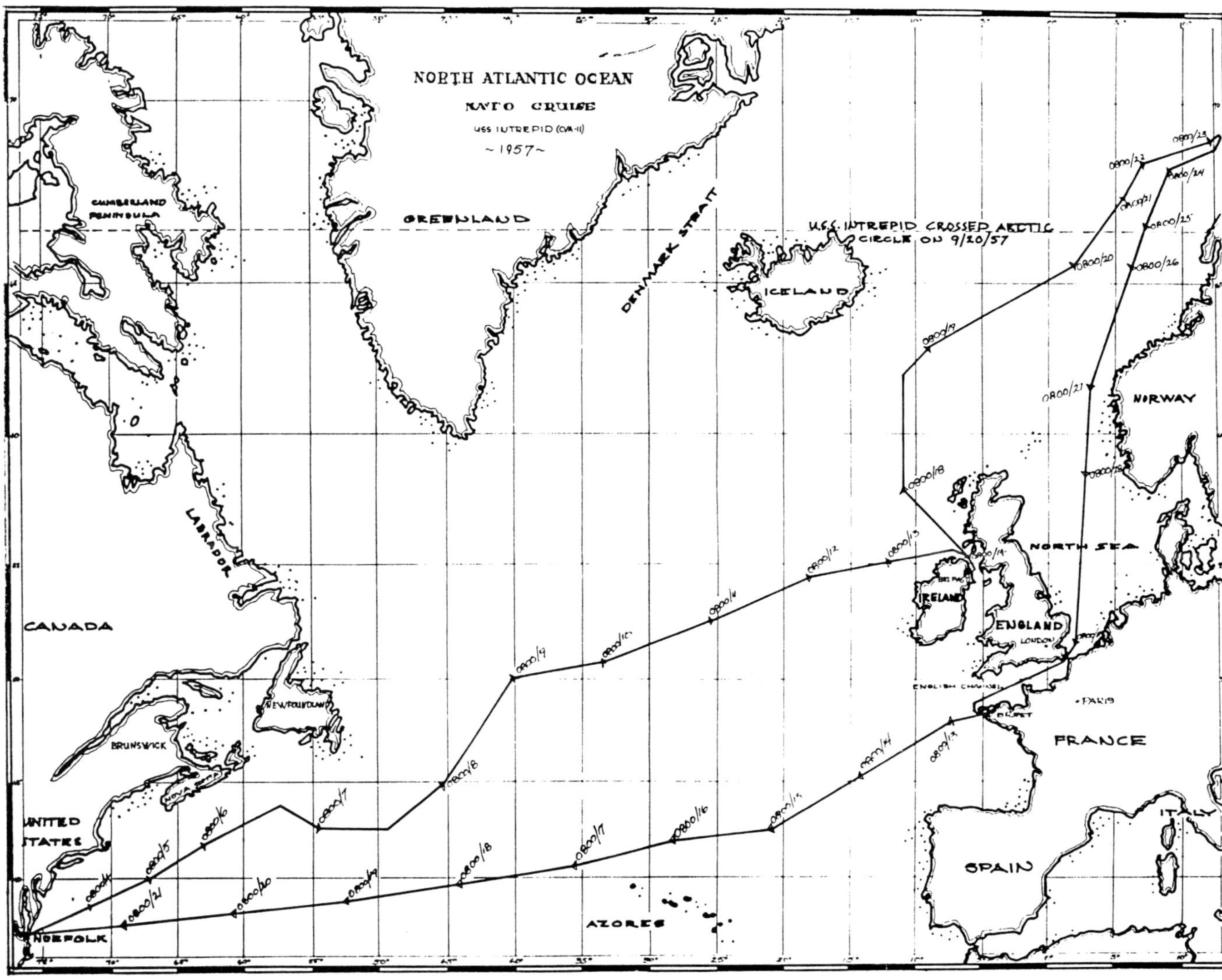

The new carrier "Cougar" planes, with tailhooks down, prepare for landing on the deck of the Intrepid. The nose probes are for mid-air fueling.

The Intrepid docked next to the new super-carrier Forrestal (CVA-59) November of 1957.

Intrepid passing under the New York Bridge in May 1963.

March 1962 for a brief refitting program. On 2 April with Air Antisubmarine Group 56 on board she went to sea for the usual shakedown cruise to test her new equipment. Her aircraft now totaled 44 in her new role (16 of them were helicopters), and her compliment was reduced to 128 officers and 2,000 men.

Intrepid was selected as the carrier to be on station for the recovery of astronaut Scott Carpenter after splashdown in his Mercury space capsule (Aurora 7) on 24 May 1962. Two helicopters were sent out from *Intrepid* and one picked up Carpenter and brought him aboard the carrier which then returned to the U.S.

CVS-11 spent the Summer of 1962 at sea in personnel training and was overhauled at Norfolk in the Fall. With her overhaul completed, *Intrepid* left for war games exercises in the Caribbean in late January 1963. This period of duty was interrupted in February as *Intrepid* joined other ships in hunting down a hijacked freighter. The mutineers were pro-Castro Communists. The vessel was tracked down and its crew surrendered at Rio de Janeiro. By 23 March *Intrepid* was back at Norfolk. She operated in waters from Maine to the Caribbean in the next year, training personnel in her antisubmarine duties. In June of 1964 she again sailed for the Mediterranean to join the Sixth Fleet in antisubmarine hunter-killer exercises. One of her duties during this deploy-

John Young and Virgil Grissom are picked up by Intrepid helicopters from their spacecraft Molly Brown after a three-orbit flight on 23 March 1965.

ment was the shadowing of a Soviet task group of ships. The "cold" war was rather warm in these years. On the way home later that Summer *Intrepid* was awarded the much sought after "E" for Battle Efficiency for her operations in the preceding year.

Intrepid sailed the U.S. east coast in the Fall of 1964, and in October she was at Yorktown to participate in ceremonies commemorating the surrender of Lord Cornwallis in 1781. The carrier then conducted training exercises for another space mission rescue assignment that took it into 1965. On 23 March 1965 *Intrepid* helicopters picked up John Young and Virgil Grissom from their spacecraft (Molly Brown) after a three-orbit flight ending in a precise and controlled re-entry. The carrier was only about 50 miles from the splash down point. The two astronauts were returned to Cape Kennedy, Florida. In April she entered the Brooklyn Navy Yard for her Fleet Rehabilitation And Modernization overhaul (FRAM II). This work was completed during the Fall and Winter and she then went on the usual shakedown cruise to Guantanamo Bay.

Assignment: Vietnam

At this time the war in Vietnam was heating up and the U.S. was deeply involved. Air cover was sorely needed around-the-clock in support of our troops in the jungles and in early 1966 *Intrepid* was ordered to duty in her old fighting grounds in the Pacific. By 15 May she was on station off Vietnam as part of Task Force 77 of the Seventh Fleet and launched Air Group 10 on their first strike against Viet Cong troop concentrations in South Vietnam. They were to fly nearly 2600 sorties from *Intrepid* sinking barges, and destroying and damaging enemy vehicles, railroad cars and bridges. During this deployment her air and deck crews established a launching record in sending off nine A-4 (jet-powered) Skyhawks and six A-1 (piston engine) Skyraiders in seven minutes with a 28 second interval between launches. On 9 October A-1 Skyraiders of Air Group 10 shot down a MiG-17 and sent another one back to his base trailing smoke. By August Air Group 10 had shifted to targets in North Vietnam. After seven months of duty in Vietnam *Intrepid* headed home and arrived at Norfolk on 23 November 1966.

CVS-11 was overhauled at Norfolk for the next several months, taken on her shakedown cruise, and by May of 1967 was on her way back to the western Pacific for a second tour of duty off Vietnam. This time *Intrepid* sailed by way of the Mediterranean and barely negotiated the Suez Canal on 1 June before it was closed down during the Arab-Israeli "seven day war." *Intrepid's* first stop was at Subic Bay in the Philippines on 15 June and a week later she was on station in the Tonkin Gulf. For the next five months she sent thousands of sorties over North Vietnam to the borders of Communist China. Air Group 10's specialty: bridge-busting. *Intrepid*

The Intrepid in Vietnam waters.

Plane preparing for take-off for strike on Vietnam.

Plane 200 on the Intrepid just prior to take-off.

A1 aircraft being readied for air-strike over Vietnam 20 May, 1966.

The Intrepid is underway in the waters of Vietnam.

The Intrepid CVS-11 with attack jet aircraft on deck, including Crusaders and Skyhawks.

An F-8 Crusader goes off Intrepid's new port side deck.

departed "Yankee Station" on 25 November 1967, stopped at Subic Bay briefly (leaving there on 30 November) and arrived home at Norfolk on 30 December. She went into the Yard on 4 January 1968 for three months of repairs and modifications.

Following the post-repair sea trials and training exercises, *Intrepid* up anchored from Norfolk for her third tour of duty off Vietnam. She arrived at Yankee Station with her destroyer escorts on 24 July and began flight operations immediately. Her mission this time was to interdict and destroy enemy supply routes in South Vietnam. Most were ground targets; occasional encounters with MiG-21s almost always ended with the Communist jets ending up second best. *Intrepid* flew off nearly 1800 combat sorties in her first 30 days in the Gulf. She then set a course for Sasebo, Japan for a brief R&R for her crew. She was back on station in the Gulf by 5 September 1968. Sortie after sortie destroyed trucks, anti-aircraft missile sites, barracks, bridges and barges. On 19 October an *Intrepid* F-8 Crusader (Lt. A. Nargi) shot down a MiG-21 and damaged a second. The carrier would be in the western Pacific until the end of December, interrupted only by brief visits to Subic Bay and Hong Kong. When bombing strikes were ordered halted on 1 November, the carrier sent its strike planes into Laos which was being used by the Viet Cong as transshipment routes. On 27 December the *Intrepid's* third tour of duty off Vietnam ended and she headed home with stops at Sidney, Australia, Wellington, New Zealand, then around Cape Horn and into the Atlantic with a stop at Rio de Janeiro, Brazil. On 8 February 1969 she steamed into Norfolk.

Intrepid CVS-11 in the Caribbean area, carrying anti-submarine aircraft. This was her final operational role.

Home Duty
Final Decommissioning

Intrepid was dry-docked on 27 February at Philadelphia for an extensive six-month overhaul and with work completed by August she was floated and sailed to her new home at the Naval Air Station, Quonset Point, Rhode Island where she was berthed on 7 September 1969. *Intrepid* steamed south to Guantanamo Bay on 13 October and there conducted training exercises for new men until 25 November. The end of 1969 saw *Intrepid* back home at Quonset Point.

For the first four months of 1970 the carrier operated in the Gulf of Mexico, conducting training exercises from Pensacola, Florida, Corpus Christi, Texas, and other Gulf ports. From May to August the carrier was at the Boston Navy Yard for overhaul, conducted her sea trials late in August and returned to Quonset Point on 3 September. *Intrepid* conducted flight trials until 16 September when she got underway for Halifax, Nova Scotia for joint Canadian-U.S. operations. Following these the carrier steamed south to Bermuda conducting anti-missile training for her Air Group. Some storm damage incurred to the carrier on her way home following these exercises and she put into Boston for repairs, but she was soon back home at Quonset Point where she spent the remainder of 1970. *Intrepid* was now the Navy's oldest operational carrier — over 27 years old.

She operated out of her home port for the first three months of 1971, honing her skills in anti-submarine warfare. On 16 April she headed east for a six month deployment in the Mediterranean Sea and eastern Atlantic Ocean. She ranged far and wide along Europe's Atlantic coast (many times being shadowed by Soviet Union long-range reconnaissance aircraft) with ports of call that included Scotland, Denmark, Norway and Germany, and England, France, Portugal and Spain. Operations off Norway took her above the Arctic Circle on several passages. 29 November 1971 saw *Intrepid* steaming into the Boston Navy Yard for a brief period of upkeep and maintenance. With this work completed by late in January 1972 she carried out training exercises for her crew for about six weeks and then fleet exercises with Portuguese and Spanish ships until her return to her Rhode Island port on 23 April. *Intrepid* sortied from Quonset Point on 11 July for a three month's cruise in north European waters with other NATO units, making port calls at the Netherlands, Denmark, Norway, England and Scotland. *Intrepid* was back home by 20 October but for barely a month; on 24 November she sailed again to join the Mediterranean fleet, with stops at Italy and Spain by year's end and later in Greece and Portugal. All during these cruises it was training, training, training for her crew. And her crew and Air Groups became very efficient at their job of hunting down submarines. She sailed for home in April of 1973 and by the 25th was again at Quonset Point, Rhode Island.

But the tough, battle-tested carrier was getting old. She was now approaching her 30th year of service to her country. On 23 July *Intrepid* was classified as "In Commission in Reserve" and preservation work began to prepare her for joining the reserve (mothball) fleet. The new nuclear-powered carriers were bigger, faster, and carried a heavier punch. Finally, *Intrepid* was decommissioned for the third and last time on 15 March, 1974. Eventually, everyone was sure, she would be headed for the scrap pile; it was only a matter of time. From about 1954 *Intrepid* was commanded by a series of skippers, each one in command for about one year. By decommissioning in 1974, 30 able men had commanded *Intrepid* in as many years.

The Intrepid in Philadelphia to be a part of the Navy-Marine Corps Bicentennial of 1975-1976. Main mast was removed to get under low bridges on the way to dock.

A briefing on the flight deck of the Intrepid.

The Intrepid offers a greeting on a visit to Quebec.

The Intrepid thanks Quebec as it departs it's port.

A Lady is Saved

But her service was not yet ended. On 13 October 1975, by act of Congress, *Intrepid* was designated as the official U.S. Navy and Marine Corps Bicentennial Exposition Vessel at Philadelphia in celebration of the nation's and the Navy's 200th birthday. Thousands upon thousands visited the gallant carrier in the ensuing year, but by the end of 1976, the scrap heap loomed again. But then, a miracle.

The following paragraphs are taken directly from an article written by John Culhane titled "The Man Who Bought an Aircraft Carrier" which appeared in the April 1990 issue of Reader's Digest. It is the story of the saving of the *Intrepid* and the man who did it.

> Despite her glorious past, the *U.S.S. Intrepid* was condemned to the scrap heap in 1976 — an obsolete aircraft carrier in a modern, state-of-the-art Navy. Then a quiet patriot named Zachary Fisher came to the rescue. This prominent New York City builder was indignant. *How can there be such a dismal ending for such a great ship?* he asked himself. *How can we cut up our own history for razor blades?*
>
> Zachary Fisher likes to say that he was born with Naval aviation — in 1910 when the first plane took off from the *U.S.S. Birmingham.* At age 16, he left high school without finishing to help his father and older brothers in the family construction business. He laid bricks, earned his union card and by age 19 was supervising bricklayers.
>
> Fisher's knee was severely injured in a construction accident, however, and when Japan attacked Pearl Harbor on December 7, 1941, he was rejected by the military, dashing forever his hopes of serving his country in the war. Searching for other ways to be of use, he helped the U.S. Army Corps of Engineers build coastal fortifications.

Mr. & Mrs. Zachary (Elizabeth) Fisher founder and Chairman of the Board for the Intrepid Museum Foundation. (Photo courtesy of Intrepid Sea Air Space Museum)

The family's business continued to prosper after the war. Fisher and his wife, Elizabeth, became active in work for wounded veterans. They helped finance the Veterans Bedside Network, which afforded hospitalized veterans the chance to write, direct and act in their own shows and broadcast them from one veterans facility to another.

It was Fisher's growing reputation as a man who remembered while others forgot that led to an invitation to a power breakfast of 20 prominent businessmen on November 30, 1978. Larry Sowinski, author and Naval history buff, told them of the impending fate of the *Intrepid.* He believed that the ship could be transformed into an aerospace and Naval museum for New York, with historical and technological exhibits. Watching grainy black-and-white newsreel footage of the *Intrepid* in action, Fisher recalled when young men died in infernos of smoke and fire and what their courage had meant to those back home.

Can I possibly take on such a big project? he wondered. *I'm 68 years old. Can I see it through? God knows, I want to. I've made a lot of money in this country. I'd like to give some of it back.*

Sowinski took Fisher to Philadelphia, where the *Intrepid* had been rusting away since the bicentennial. The only way to board her was up a steep, narrow ramp. Fisher's bad leg couldn't make it, so Sowinski, a gentle bear of a man, carried the builder up piggyback.

Fisher's heart sank as he got on deck. All around him was peeling paint, weather damage, neglect.

I am a builder, he said to himself. He threw his shoulders back and imagined the *Intrepid,* those priceless seven full decks of living history, against the magical skyline of Manhattan. *I will save the Intrepid,* he promised himself. *There will be no turning back.*

The first problem was to get the ship off Navy rolls. A Navy aircraft carrier had never been transferred to a private foundation before, and Fisher had to show financial capability. His personal fortune, estimated by *Forbes* magazine as among the nation's top 400, passed that test. But an act of Congress would have to authorize the transfer.

On February 27, 1979, the Intrepid Museum Foundation was created under Fisher's leadership as a not-for-profit corporation. Congress acted 26 months later to transfer the *Intrepid* to the foundation.

Another challenge lurked in the building regulations of the City of New York. "*The codes treated the Intrepid like a building laid on its side in the water,*" recalls Fisher with a smile, "*That doesn't work.*" The city didn't, for instance, have a fire rating for a ship's steel bulkhead.

Beyond private donations from the Fishers and others, the museum was to be financed by the sale of bonds and a loan from the city. But no bond could be written until the foundation had its plans in place. And it couldn't do that until New York's building codes were satisfied.

Fortunately, then Mayor Edward Koch saw the *Intrepid* as a centerpiece of the city's planned West Side development. He ordered the restoration of Pier 86 on the Hudson River as the ship's new berth.

Eventually, New York City agreed to safety modifications so stringent that they serve as guidelines for other vessels used as moored structures. In January 1982 the foundation sold $14.2 million in tax-exempt bonds. A federal loan, obtained through the City of New York, provided an additional $4.5 million.

That February the *Intrepid* was towed from Philadelphia to Bethlehem's Hoboken Ship Yard in New Jersey for restoration. The modifications proved so extensive that the ship missed its scheduled early summer opening and the Fourth of July crowds that Fisher and the foundation had counted on. Worse, there were tremendous cost overruns.

On June 12, Zachary and Elizabeth Fisher and Larry Sowinski boarded the vessel in Bayonne, N.J., for the last cruise of the *U.S.S. Intrepid.* It took 6 and 1/2 hours in heavy rain to navigate six nautical miles through the Upper Bay and into the Hudson estuary to the pier in midtown Manhattan, but all three wanted to be part of that final voyage. The *Intrepid,* Fisher said, had "*become my life.*"

The Intrepid departs Philadelphia for the Bethlehem's Hoboken Ship Yard in Bayonne, New Jersey for restoration. (Photo courtesy of Intrepid Sea Air Space Museum)

Intrepid's restoration is underway at the Bethlehem's Hoboken Ship Yard. (Photo courtesy of Intrepid Sea Air Space Museum)

Now they had seven weeks to install the exhibits if they were going to make an early August opening. There was no electric power when they first got there, so they had to operate on portable generators. Then the exhibit company went on strike. Through it all, Fisher seemed unperturbed, "*like the good captain in heavy seas*," said Sowinski.

Opening day for *Intrepid* Sea Air Space Museum was August 4, 1982. Fisher remembers it this way: "*I watched kids in sneakers wandering in awe all over this piece of history. I wondered if any of them would be inspired to do something for their country.*"

Several weeks later, he got the perfect example of what a visit to a historic ship can do. George Bush, then Vice President of the United States, flew to New York to dedicate *Intrepid* Hall, Honoring America's wartime role in the Pacific. Bush spoke of his own excitement as a 12 year old when he had visited a Navy destroyer anchored in the Hudson. That visit had inspired him to join the Navy, Bush said.

A first-year attendance of one million visitors had been projected for the museum. But only 780,000 came, and that first fiscal year ended with a deficit of $3.5 million. Quietly, Fisher kept paying the bills. He told the New York *Times*: "*This museum will never close.*"

The Intrepid Sea, Air and Space Museum as it looked in 1986. (Photo courtesy of Intrepid Sea Air Space Museum)

Fisher renegotiated the loan from the city, got help from local businesses and continued to pour his own money into the project. With his help, the museum's revenues were enough to pay operating expenses. But there were lean times. In the fiscal year ending in April 1985, the museum operated at a loss of just over $1 million. And that was on top of $2.7 million in interest payments made by Fisher.

After attendance picked up in 1985, however, Fisher could mount the flight deck and declare his confidence that the museum would survive. His hopes were fixed on the summer of 1986, when the rededication of the Statue of Liberty would bring millions of visitors to New York. Fortune smiled early, when the *Intrepid* was designated a National Historic Landmark. She participated in the 100th birthday celebration of the Statue, and the opening of the Jacob K. Javits Convention Center in the neighborhood also helped.

But it was the growing number of kids who gave Fisher his greatest pleasure. "*She's a symbol,*" said one 12 year old visitor. "*But she's also the real thing, right?*"

Fisher nodded, and pointed in the direction of the Statue of Liberty. "*That's right, son. We're the Fighting Lady of the Harbor — the one that helped keep Miss Liberty's torch lit.*"

He watched the children in Pioneers Hall, learning about the first shipboard takeoff from the *U.S.S. Birmingham.* He followed them to Technologies Hall, where they touched a space capsule.

In Navy Hall, he picked out a teen-age boy and girl watching a wide-screen film he had helped produce about flight operations aboard the nuclear carrier *U.S.S. Nimitz.* The teen-agers glowed with excitement at this dramatic airborne marriage of technology and human dedication. Then they grew sober-faced during a hand-in-hand tour of the Hall of Honor for Congressional Medal of Honor recipients — those Americans recognized for valor, above and beyond the call of duty.

Dan Rather on CBS News once watched children explore the *Intrepid* and was so moved that he did a commentary in the ship's voice:

Anthony Zollo's two grandsons Shawn (left) and Michael (right) stand in front of a model of the first Intrepid on Intrepid's hanger deck in 1992.

The Gemini Space Capsule on display in Technology Hall. (Photo courtesy of Intrepid Sea Air Space Museum)

> *"They sent me off with 3000 boys —farm boys and drug-store clerks, now fliers and sailors. Boys don't have majesty, but my boys did.*
>
> *"You there, child of today, standing on my deck in your brand new sneakers, a boy fell there. He manned a five inch gun, and he fell. He fell where your shadow is. Walk softly there.*
>
> *"I am the Intrepid. I survived. History lived within me. He lives within me still."*

The men who served on the *Intrepid* also come to visit. Men like Alonzo A. Swann, Jr., of Gary, Ind., now 64, see photographs of action that took place on October 29, 1944.

The exhibit tells the story of a Japanese kamikaze attack. Swann and 21 others manning the 20 mm anti-aircraft guns at Gun Tub 10 steadfastly shot off the suicide plane's wing and tail. The pilot released his bomb prematurely into the sea, but the plane kept coming. It flew across the flight deck, and the remaining wing hit Gun Tub 10, splashing the crew with aircraft fuel that flashed into searing flame. Ten died in the attack. Swann himself suffered severe burns and shrapnel wounds, but he lived to face further attacks in 1944 and 1945. After the Japanese surrender, the *Intrepid* brought back not only its triumphant crew, but many grateful POWs.

Former Marines of the 7th Detachment of Intrepid 1943 to the end of WW II. (l. to r.): Ray Dow, Anthony Zollo, Sr., Col. Keagy, George Cramer and Mr. Haase. Photo taken August 1992.

Swann is making plans to tour the ship with his son and grandson. He will show them the plaque that tells of the heroism of Gun Tub 10 and with a tender finger trace his own name and the names of his fallen shipmates.

The *Intrepid* is safe now. And Fisher's dream is realized. *For years to come,* he thought recently, watching the flag rise above the ship's bridge and seeing the Empire State Building shining in the distance, *long after I'm gone, this Fighting Lady will be telling her stories of the meaning of freedom. She'll be telling our children's children of duty, honor, faith. Of hope and courage and pride. Of compassion and sacrifice and endurance. The old values that have been at the heart of America from the very beginning.*

Intrepid was officially stricken from the Navy's register of vessels on 23 February 1982 and was transferred from the Navy to the Intrepid Museum Foundation. The big carrier is now the focus of the **Intrepid Sea Air Space Museum** moored at Pier 86, port side, on the Hudson River, New York City, at West 46th Street and 12th Avenue. Moored opposite the *Intrepid* are other ships on display including the missile submarine *Growler*, the destroyer *Edson,* and the 1936 Coast Guard lightship *Nantucket*. Since the museum's mission is to present and explain significant 20th century history and technology with emphasis on sea, air and space accomplishments, many of its exhibits deal with the history of the carrier itself and the role of the Navy in the great historical events of this century. But also there are exhibits by the Army, a World War I exhibit, a space capsule display, a Congressional Medal of Honor display, many fully restored rooms and spaces on the ship, and on display are more than 70 aircraft, both from World War II and very recent. The displays are both on the pier, and in several "Halls" on board the *Intrepid*, including Technologies Hall, Pioneer Hall, Intrepid Hall, United States Navy Hall and of course the expansive Flight Deck of the carrier itself. Numerous slide and video shows are constantly on view for visitors to the Museum. Other exhibits are temporary and rotate from time to time.

On 21 April 1991, I visited the *Intrepid* after forty-six years. She was the carrier I served on and she went on for years. She was mothballed twice and made three tours of duty (three years) at Vietnam. She was the carrier which picked up astronauts in the Pacific after the men landed back on earth. One of the outer space

A modern battleship salutes the "Senior Citizen of the Seas" during Fleet Week 1991 Desert Storm Welcome Home. (Photo courtesy of Intrepid Sea Air Space Museum)

"Pioneer's Hall" Exhibit. (Photo courtesy of Intrepid Sea Air Space Museum)

"Intrepid Hall" Exhibit. (Photo courtesy of Intrepid Sea Air Space Museum)

"Navy Hall Exhibit". (Photo courtesy of Intrepid Sea Air Space Museum)

vehicles was named The Intrepid Module. As I write about these old memories and thoughts and as I read of the exploits of so many men and so many ships at sea, I say to myself, I am so proud to have been part of it all.

I shall never forget the heroism, the blood and guts of all the men, the adventure, the sights of men and ships at war, the days and nights at sea twelve thousand miles from home. This magnificent ship, the *U.S.S. Intrepid,* the most hit ship in the Pacific in World War II is still afloat today, almost fifty years after the keel was laid which was December 1, 1941 and commissioned on August 16, 1943. The carrier *Intrepid* is now a floating museum in New York. She is tied up at a Pier known as Intrepid Square at the intersection of West 46th Street and 12th Avenue in New York.

As I parked my car in the lot I paused and took a good long look at this lady of the sea with so much history behind her. I did not get out of the car for some ten minutes as I looked and swept my eyes all over her full eight hundred eighty six feet. My mind opened and out flooded so many memories of the years past. The ship has undergone changes on at least three occasions and does not look as she did when she was brand spanking new when I went on board her in August of 1943. She looks tired now, her paint is sadly in need of refinishing. Her sides have hundreds of dents. Her superstructure is greatly

The CIC Exhibit. (Photo courtesy of Intrepid Sea Air Space Museum)

The A-12 Blackbird on display at Intrepid. (Photo courtesy of Intrepid Sea Air Space Museum)

changed and looks nothing like the ship I left in 1945. But her appearance is not bad; it is the style and shape of today's ships.

Gone are the many 20 mm guns along her sides and gone are the 40 mm guns that were positioned at the front of each of the two dual 5/38 mounts. I looked up at the spot even with the signal bridge of the island structure where my 20 mm gun used to be in Tub #3, starboard side, outboard of the island - gun #1 of five. I stared at the spot for a long time and the times of battle came to mind and I remembered those times as if it were yesterday. I looked up at the flight deck and also remembered the ten thousand flights taking off and landing. The roar of seventy and sometimes eighty plane engines warming up. The memories of the takeoffs, every one and a half minutes or so in emergencies. The wonder of whether the plane will lift as it leaves the deck or will the engine sputter and the plane hit the water. On landing you wonder: will the tail hook hold, is the pilot O.K. or is he injured? Will the first cable stop him or will he be stopped by the second or third cable? If the hook pulls out or misses the deck cable, will the first or second or third cable barrier stop him? Will all planes return to the ship or will some have been lost on this strike? Will some ditch in the water due to low fuel or under-carriage damage or an injured pilot?

You think of the day the Japanese Betty bomber came in on the starboard side directly in line with your gun, flying only seventy or eighty feet above the water, and you say to yourself — I've got to hit this guy before he hits me. And you do hit him and he explodes so close to you that you become scared as hell. It is then that the gunnery officer or someone else on the bridge will holler down to you "Nice shooting!" And the Captain may even later on look you up and say in person to you, "Well done." You feel so proud that you almost want to cry.

I got out of the car and walked the full length of the pier and the ship, and then I went on board. I was excited now and wondering what I will see and who is on board her. As I bought my ticket for admission to the ship, without thinking I made the remark to the ticket seller, "I have come home." He looked puzzled and then I told him I am a former crew member of forty-seven years ago and he replies, "Welcome aboard." I walked up the stationary steel stairway that has been built to get aboard the ship. I stepped off the steel platform onto the ship and after a half dozen steps I stopped and looked around. I am on the hanger deck amidships on the port side. The carrier is nosed into the pier with her bow facing the West 46th Street intersection with 12th Avenue, or better known as Intrepid Square. An older man, possibly my age, walked over and before he could ask the questions he asks a lot of people who come on board, I said, "I'm a former crew member back at commission in August 1943 and I have come to get the plank that I own on this ship." His eyes lit up and his question was going to be "Are you a former crew member?" We shook hands and introduced ourselves to each other and he invited me to his office across the deck on the starboard side, amidships.

We talked and it wasn't five minutes before he told me he was one of the original crew members also and that he had been an airedale (plane handler). The man has a bad leg and walks with a cane. It seems that he was in the wrong place at the wrong time that fateful night at eleven minutes after midnight, early on February 17, 1944 when a Japanese Betty bomber, flying low on the water came upon the wake of the *Intrepid.* He dropped a torpedo and swung to the right and up the starboard side, still low on the water. The torpedo hit under the stern below the waterline about fifteen feet, and exploded, leaving a large hole in the hull of the ship on the starboard aft area. The torpedo penetrated into the ship's interior killing men and knocking out the rudder. The man's name was Frank Doria; a pleasant, nice man I thought, as I got to know him a little better. We talked about a hundred different things and about the battles of the *Intrepid* and he just could not get enough information. So I told him he would get a copy of this book personally from me when it is finished. (We have become such good friends that we talk on the phone now at least once a month).

Aerial view of the Island. (Photo courtesy of Intrepid Sea Air Space Museum)

It has been nearly 48 years now since *Intrepid* last launched her fighters and bombers against our enemy of World War II. She has sent her planes against other enemies since then, and in over 30 years of service the gallant ship never backed down — never retreated — from a fight. In the final analysis, of course, these are not the actions or qualities of steel hulls and wooden flight decks; these are qualities of her officers and men. It is to these we are forever indebted.

The Intrepid at her berth in 1991 showing the New York City skyline.

References

Kurzman, Dan. Fatal Voyage: The Sinking of the Cruiser U.S.S. Indianapolis. McMillan Publishing Co., New York, 1990.

Morison, Samuel Eliot. History of United States Naval Operations in W.W. II, Vol. XII. Little Brown and Co., Boston, Mass., 1958.

Belote, James H. and William M. Belote. Typhoon of Steel. The Battle for Okinawa. Harper and Row, Publishers, New York, Evanston, and London.

McCormick, John. The Right Kind of War. Naval Institute Press, Annapolis, Maryland.

Potter, E.B. Admiral Arleigh Burke. A Biography. Naval Institute Press, Annapolis, Maryland (Random House, N.Y.).

Potter, E.B. Nimitz. Naval Institute Press, Annapolis, Maryland.

Potter, E.B. Bull Halsey. Naval Institute Press, Annapolis, Maryland.

Miller, Nathan. The Naval Air War, 1939-1945. Nautical and Aviation Publishing Co. of America, 1980. (Reprinted by Naval Institute Press, 1991).

Monsarrat, John. Angel on the Yard Arm: The Beginnings of Fleet Radar Defense and the Kamikaze Threat. Naval War College Press, Newport, R.I., 1985.

Stewart, William H. Ghost Fleet of the Truk Lagoon. An Account of "Operation Hailstone," February, 1944. Pictorial Histories Publishing Co., Missoula, Montana.

Jensen, Lt. Oliver. Carrier War. Pocket Books, Inc., New York, 1945.

Silverstone, Paul H. U.S. Warships of World War II. Naval Institute Press, Annapolis, Maryland, Ian Allan Ltd., London, 1965. Reprinted 1989.

Terzibaschitsch, Stefan. Aircraft Carriers of the U.S. Navy. Conway Maritime Press, Greenwich, London. 1978.

Polmar, Norman. Aircraft Carriers. Doubleday & Company, Garden City, New York. 1969.

Sumrall, Robert F. Warship's Data. 4. U.S.S. Intrepid (CV-11). Pictorial Histories Publishing Company, Missoula, Montana, 1989.

In additioin to the above, I have used numerous magazine and newspaper sources for information, including:

Time magazine, especially issues of Nov. 23, 1942, July 10, 1944, and Sept. 11, 1950.

U.S. Naval Institute Proceedings Magazine, July, 1956: Japanese Parade of Ships in Tokyo Bay.

Newsweek magazine, especially the issue of May 3, 1943.

U.S. Naval Institute Proceedings, June 1949.

Bureau of Naval Personnel Information Bulletins (All Hands) of April and August, 1945.

Intrepid's monthly news releases which I had retained in my possession for all those years.

Appendix

The signal gang at work 4 February as the ship moves out of Majuro Atoll. (National Archives Photo)

I. Muster Roll of Officers and Enlisted Men of the U.S. Marine Corps on the USS Intrepid

From 10 August to 31 August, 1943

Captain
Hammond, Bruce B.

Second Lieutenants
Keagy, Lowell T.
Nickell, Louis G. III

First Sergeant
Duncavage, Anthony

Gunnery Sergeant
Crenshaw, Harry K.

Platoon Sergeants
Daus, Bernard L.
Jones, Norman P.

Sergeants
Franklin, Lawrence C.
Grund, August C.

Corporals
Alexander, Donald J.
Dickson, William Y.
Gettig, Spencer T., Jr.
Heald, Chester G.
Marutiak, John S.
Wilson, William H.

Privates First Class
Balentine, Robert W.
Fabia, Robert M.
Himes, William C.
Plumm, Oscar H.
Sledjeski, Frank A.
Vojtko, George R.

Field Music First Class
Cresta, John D.

Private
Smith, Nevin

Marine Corps Reserves Class III

Sergeants
Kelly, Paul B.
Weldon, Arthur J.

Corporals
Langan, James J.
Lesnett, Dean M.

Assistant Cooks
Kelley, Floyd G.
Santacroce, Orlando A.

Privates First Class
Bedick, John Jr.
Black, Albert J.
Chopie, George
Conko, Edward J.
Cramer, George P.
De Walt, Maurice P.
Dow, Raymond D.
Dugan, Patrick H.
Eddy, Frederick W.
Gaither, Charles E.
Guthrie, James C.
Hartmann, Albert J.
Landes, Stanley A.
Mc Hugh, Emanual J.
Puskas, George B.
Smeeding, Richard C.
Welker, Charles F.
Wicker, Arthur J., Jr.
Yanuzzi, Joseph W.

Field Music First Class
Zjawin, Edward M.

Privates
Abbott, Kenneth G.
Baden, Leonard
Beale, William H.
Benkowski, Stanley J.
Bowen, Philip S.
Bush, James W.
Butcher, Thomas H.
Chiapparelli, Louis J.
Duffy, Robert H.
Guertin, Ernest O.
Gyenge, Ernest L.
Haese, Francis R.
Halter, John
Hannan, Thomas E.
Jackson, Andrew S.
Keith, Samuel B.
Kelly, Jack L.
Kloby, John W.
Kovach, Joseph T.
Mentus, John
Mesi, Pasquale P.
Monaghan, Robert E.
Murphy, Gerald T.
Pretola, Vincent
Saliba, David J.
Sardo, Ross A.
Sawran, John
Smithson, William R., Jr.
Stewart, Raymond E.
Wicks, Edgar E.
Wiley, William C.
*Zollo, Anthony F.

From 1 December to 31 December, 1943

Captain
Hammond, Bruce B.

First Lieutenants
Keagy, Lowell T.

Second Lieutenants
Nickell, Louis G. III

First Sergeant
Duncavage, Anthony

Gunnery Sergeant
Crenshaw, Harry K.

Platoon Sergeants
Daus, Bernard L.
Jones, Norman P.

Sergeants
Franklin, Lawrence C.
Gettig, Spencer T., Jr.
Grund, August C.

Corporals
Alexander, Donald J.
Heald, Chester G.
Marutiak, John S.

Privates First Class
Balentine, Robert W.
Fabia, Robert M.
Himes, William C.
Plumm, Oscar H.
Sledjeski, Frank A.
Smith, Nevin

Field Music First Class
Cresta, John D.

Marine Corps Reserves Class III

Platoon Sergeants
Weldon, Arthur J.

Sergeants
Kelly, Paul B.

Corporals
Dugan, Patrick H.
Mc Hugh, Emanual J.
Langan, James J.
Lesnett, Dean M.

Assistant Cooks
Kelley, Floyd G.
Santacroce, Orlando A.

Privates First Class
Abbott, Kenneth G.
Bedick, John Jr.
Benkowski, Stanley J.
Beale, William H.
Black, Albert J.
Bowen, Philip S.
Bush, James W.
Butcher, Thomas H.
Chopie, George
Cifelli, August P.B.
Conko, Edward J.
Cramer, George P.
De Walt, Maurice P.
Dobrolet, Walter J.
Dow, Raymond D.
Duffy, Robert H.
Eddy, Frederick W.
Finney, Ira A.
Gaither, Charles E.
Grube, Dale O.
Guertin, Ernest O.
Guthrie, James C.
Gyenge, Ernest L.
Halter, John

Hannan, Thomas E.
Hartmann, Albert J.
Keith, Samuel B.
Kovach, Joseph T.
Landes, Stanley A.
Mentus, John
Mesi, Pasquale P.
Monaghan, Robert E.
Page, Maxwell L.
Principe, James
Puskas, George B.
Saliba, David J.
Sardo, Ross A.
Sawran, John
Smeeding, Richard C.
Smithson, William R., Jr.
Welker, Charles F.
Wicker, Arthur J., Jr.
Wiley, William C.
Zollo, Anthony F.

Field Music First Class
Selvaggo, Francis P.

Privates
Baden, Leonard
Haese, Francis R.
Jackson, Andrew S.
Kloby, John W.
Murphy, Gerald T.
Pretola, Vincent
Stewart, Raymond E.
Wicks, Edgar E.

Transferred Corporals
Dickson, William Y.
Vojtko, George R.

Privates First Class
Chiapparelli, Louis J.
Yanuzzi, Joseph W.

From 1 December to 31 December, 1944

Captain
Hammond, Bruce B.

First Lieutenants
Keagy, Lowell T.
Nickell, Louis G. III

First Sergeant
Duncavage, Anthony

Gunnery Sergeant
Crenshaw, Harry K.
Daus, Bernard L.

Platoon Sergeants
Jones, Norman P.

Sergeants
Gettig, Spencer T., Jr.
Grund, August C.
Heald, Chester G.
Smith, Joel S.

Corporals
Balentine, Robert W.
Bilek, Joseph
Grube, Dale O.
Sledjeski, Frank A.

Privates First Class
Bell, Milton R.
Burk, William H.
Himes, William C.
Smith, Nevin
Somerville, George A.

Private
Turley, Herbert E.

Field Music Corporal
Cresta, John D.

Marine Corps Reserves Class III

Platoon Sergeant
Weldon, Arthur J.

Sergeants
Kelly, Paul B.
Page, Maxwell L.

Corporals
Bedick, John Jr.
Conko, Edward J.
Estes, Jean F.
Jongeneel, Clifford
Mc Hugh, Emanual J.
Langan, James J.
Lesnett, Dean M.
Smeeding, Richard C.
Smithson, William R., Jr.

Assistant Cooks
Gyenge, Ernest L.
Santacroce, Orlando A.

Privates First Class
Baden, Leonard
Beale, William H.
Benkowski, Stanley J.
Beringhele, Joe M.
Butcher, Thomas H.
Cifelli, August P.B.
Collins, Lawrence D.
Cramer, George P.
De Walt, Maurice P.
Dobrolet, Walter J.
Dow, Raymond D.
Duffy, Robert H.
Eddy, Frederick W.
Forish, George L.
Halter, Frank T.
Halter, John
Hannan, Thomas E.
Hartmann, Albert J.
Henley, Fred E.
Hoppy, Lawrence E.
Jackson, Corval R.
Jenkins, Stuart A.
Johnston, Walter G., Jr.
Keith, Samuel B.
Klein, Kenneth H.
Krusz, Walter J.
Mentus, John
Milligan, Jack H.
Monaghan, Robert E.
Morse, Glenn J.
Parsons, Roy H.
Pretola, Vincent
Principe, James
Puskas, George B.
Quimby, Edward C.
Rydzefski, Joseph F.
Sardo, Ross A.
Sawran, John
Welker, Charles F.
Wicker, Arthur J., Jr.
* Zollo, Anthony F.

Privates
Pope, Earl D.
Ross, William M.
Stout, Ward W.

Transferred Privates First Class
Deringer, John A.
Meyers, Norman H.
Pearmine, Lester I., Jr.

From 1 December to 31 December, 1945

Captain
Davis, Bert, Jr.

First Sergeant
Tobin, John P.

Gunnery Sergeant
Crenshaw, Harry K.

Platoon Sergeants
Heald, Chester G.
Krueger, Harold A.

Sergeants
Bell, Milton R.
Bilek, Joseph

Corporals
Hardman, Hubert E.
Irvin, Raymond F.
Somerville, George A.
Turley, Herbert E.

Privates First Class
Blough, Richard
Chenault, James D.
Edwards, Charles
Fleming, Arch E.
Fuglestad, Marvin S.
Gibson, Clyde I.
Greene, Donald F.
Hale, Bobby D.
Henderson, Conrad L.
Hendrix, Willie B.
Herrington, Harlin A.
Holman, Milbert E.
Hood, William K. Jr.
Horton, Lemuel D.
Hudgins, Talmadge H.
Hundshamer, Francis X.
Janke, Albert A.
Portney, John L.
Sowada, Wilfred B.
Thomas, Charles D.

Winters, August J., Jr.
Yeaton, Roland A.
Young, Thomas C.

Private
Blume, Paul C.
Brestal, James
Delay, Ralph E.
Hanke, Harvey E.
Ogden, Arthur R.
Rhoads, Roy D.

Marine Corps Reserves Class III

Second Lieutenant
Ter Horst, Jerald F.

Corporals
Doolittle, Wilfred H., Jr.
Ross, William M.
Stout, Ward W.

Privates First Class
Darney, Leo R.
Dimmick, George C.
Dirom, Robert A., Jr.
Eads, William M., Jr.
Glenicki, Edward T., Jr.
Grandstaff, Walter B.
Laurence, Richard S.
Nelson, Richard M.
Newlin, Gerald F.
Piccione, Philip
Pope, Earl D.
Rasmussen, Elmer C., Jr.
Smith, Clifford A.

Privates
Allen, Corbin E.
Bagley, Louis C.
Bauer, Ralph C., Jr.
Haman, Raymond W.
Mullen, James
Perhus, Philip H.

The flagship Intrepid leads the parade in October of 1944. (National Archives Photo)

II. The Men Who Gave Their Lives for the U.S.S. Intrepid CV-11

World War II

Ship's Company

Raymond F. Adkins, S2C
Oswald S. Bauer, S1C
Royal H. Bender, S2C
James C. Bracken, Sr., S2C
Leander Brewer, Jr., STM1C
James E. Caskins, S1C
Charles H. Casteel, S1C
Alfonso Chavarrias, GM3C
Joseph W. Cody, ST3C
George B. Colwell, S2C
Clyde C. Compton, S1C
Odell H. Cooke, S1C
Hugh A. Cooper, S1C
Lawrence J. Covell Jr., S2C
Elton L. Darden, S1C
Henry F. Darnauer, S1C
Russell E. Davis, S1C
Edward Davison, STM1C
Louis N. Defedele, S1C
David Dermenegildo, S2C
Lt. Donald D. Dimarzo
Raymond H. Doulette, S1C
Willie E. Dousay, Jr., S1C
James M. Duren, S1C
Harold C. Ellen, AOM3C
Everett J. Eure, EM3C
Vito T. Fiore, S1C
John F. Fischer, S1C
Grandville Fleming, STM1C
Emil V. Foletta, S1C
LTJG. Howard P. Frank
Samuel W. Grant, STM1C
ENS. John B. Gattey
John W. Giddens, ST3C
Elmer P. Graves, ACM
Wallace D. Greig, S2C
Will C. Hamblin, Jr., AM2C
John T. Hanke, S1C
Alexander F. Hantosh, S1C
Nathan B. Hecker, ACRT
Loumas E. Helms, S1C
CARP, Samuel L. Hitt
Elden S. Hjelmeng, ACOM
Raymond Hope, S1C
Delmar L. Huff, ARM1C
Donald C. Hurd, TM3C
Leslie E. Jones, RT3C
Rhett Jones, S1C
Vernon F. Kahler, EM3C
Dean R. Krouch, S2C
Robert B. Kuhns, TM3C
Harold R. Kurtz, S1C
Delmer B. Lasater, PHM3C
William H. Lear, EM1C
Jay M. McDonald, S1C
Claude E. Maddox, AOM1C
Joseph Mancini, F1C
Merl M. Martin, S1C
James Monroe, S1C
Robert P. Morgan, S1C
Alphonse V. Moscaritolo, S1C
William H. Mottaz, HA1C
Sern C. Nelson, AMM1C
Richard L. Nieto, RDM3C
Robert H. Norris, BM2C
Mike L. Oritz, AOM2C
John Paksi, GM3C
Cecil D. Peterson, S1C
Raymond A. Ray, S2C
Laurence J. Robinson, RDM3C
Raymond J. Rucinski, COX
Robert L. Schug, MM2C
Burhl A. Schultz, S1C
Walter M. Schulz, S1C
Horace J. Sides, S1C
John Z. Sliger, F2C
Fred L. Smith, STM1C
William D. Smith, PHM2C
Earl E. Sorenson, S1C
Robert M. Stallins, F1C
Eskew R. Steffens, S1C
Kenneth C. Steffens, S1C
Lee E. Stull, AOM3C
Archie L. Svec, CWT
Edward F. Tarvainis, AMM2C
James Taylor, STM1C
Robert L. Taylor, FC2C
Harold L. Tucker, S2C
Norvel L. Van Every, S2C
Reginal E. Wallace, ACMM
John B. Ward, PHM3C
Albert J. Watson, AMM1C
Gilbert L. Weiss, S1C
Harry E. Wells, S2C
George C. Whatley, Jr., S1C
Albert M. White, S2C
Harold V. White, S1C
Ralph A. Whittington, S1C
Lyle W. Willey, S1C
Earl E. Williams, S2C
James A. Winslow, PHM2C
LT. Henry H. Whitmer
Charles W. Yoder, RDM3C
Harold L. Zieg, RDM3C

Air Group 8

Kenneth W. Johann, AP1C
ENS. William A. Preston

Air Group 6

LTJG. Dale A. Barrows
LT. James E. Bridges
Robert E. Bruton, AM1C
Gilbert J. Farmer, ACRM
LT. Robert E. Gardner, Jr.
James A. Green, ARM2C
ENS. Thomas A. Hall
David E. Jensen, ARM1C
Harold F. Leach, ARM1C
John F. McCormack, ARM1C
LTJG. Robert W. Neel, Jr.
ENS. John R. Ogg
LCDR. John L. Phillips, Jr.
LT. John P. Phillips
LT. Paul E. Tepas
LTJG. Thomas M. Vaughn

Air Group 18

J.R. Adams ARM3C
G. L. Anderson ARM1C
Harry M. Bearden, Jr., ART1C
Michael J. Benak, ARM2C
William H. Besoain Jr., AOM3C
ENS. John J. Boyle
George E. Christman, Jr., AMM2C
Tom V. Collins ARM2C
Donald R. Cooper, ARM2C
Frank P. Crevoisier, ARM2C
James D. Cropper, ARM1C
Harlan R. Dickson LCMR
Rodrick L. Duchesne, ARM2C
LTJG. Ralph C. Dupont
LT. Edward H. Eisengrein
Clarence F. English, ARM2C
CDR. Mark Eslick, Jr.
William C. Ewart, ARM3C
LCDR. John W. Fish

Edward Fussell ARM3C
Bryon B. Galbreath, ARM2C
ENS. Delbert C. Goodspeed
Sterling E. Graham, ARM1C
Lt. George A. Griffith, Jr.
Lt. John W. Gruenewald
Frank L. Haynes, ARM1C
ENS. James P. Hedrick
Edward Hlywa ARM1C
Raymond V. Hopkins ARM3C
Billy G. Hunter, ARM3C
LTJG. Isaac W. Keels, Jr.
Francis R. Krantz, AOM3C
LTJG. David L. Johnson
ENS. Daniel Laner
Alfred B. Lankford, ARM2C
LTJG. Leger
LTJG. F. A. McAllister
Frederick E. McCreary, ARM1C
Lt. Wilson C. McNeil
Lt. Walter J. Madden
ENS. Harold S. Meacham
ENS. Arthur P. Mollenhauer
LTJG. John Morelle
LT. Elmer C. Namowski
LT. Fred Navas
LT. James B. Neighbours
Elmer E. Nutgrass, ARM2C
ENS. Walter Passi
Robert H. Paulson ARM2C
ENS. Piercell
LTJG. Andrew H. Rohleder, III
ENS. Clarence Rolka
Albert P. Rybarczyk, ARM2C

Andrew H. Rohleder, III from North Carolina who lost his life on 12 October, 1944.

We just get back into action and we are hit again. (National Archives Photo)

LTJG. William H. Sartwelle
LTJG. John J. Savage
Harold H. Schnack, AMM3C
Ora H. Shaminghouse Jr., AOM2C
LTJG. Raymond J. Skelly
LTJG. Ernest A. Smith
Theodore A. Smith, ARM2C
William J. Telliard, AOM2C
LT. William B. Thompson
Louis D. Vaughan, ARM3C
ENS. A. E. Watkins
ENS. Harry R. Webster, Jr.
Robert G. Westmoreland, ARM3C
Lumpkin Wood, ACM
William P. Young, ARM2C
LTJG. William C. Ziemer

Air Group 10

ENS. Ernest M. Bailey, Jr.
ENS. Donald H. Croy
ENS. Robinson W. Harris
ENS. Elmer E. Hasse
ENS. Horace W. Heath
ENS. Loren F. Isley
LTJG. Richard W. Mason
LTJG. Willard J. Miller
ENS. Willard E. Norgren
LT. Mark L. Orr
LTJG. Stanley W. Powell
ENS. William L. York

Vietnam
Ship's Company

LCDR. Richard A. Moran
ABE3 Bobby L. Spencer

Air Wing Squadrons

VA-15
CDR. LeGrande O. Cole, Jr.
CDR. Philip C. Craig
VA-106
LCDR. David F. Callahan, Jr.
CDR. Kenneth K. Knabb, Jr.
VAW-3
LT. Frederick M. Kasch
VA-66
LCDR. Edward J. Broms, Jr.
VA-176
LT. Charles A. Knochel
VAW-121
LT. Robert W. Harris

III. Intrepid has Earned the Following Battle Stars and Awards

Navy Unit Commendation
12 June - 8 December 1967

American Campaign Medal

Asiatic-Pacific Campaign Medal

1 Star/Marshall Islands Operation
Occupation of Kwajalein and Majuro Atolls: 29 January - 8 February 1944

1 Star/Asiatic-Pacific Raids: 1944
Truk Attack: 16-17 February 1944

1 Star/ Western Caroline Islands Operation
Capture and Occupation of Southern Palau Islands: 6-14 October 1944
Assaults on the Philippine Islands: 9-24 September 1944

1Star/Leyte Operation
3rd Fleet supporting operations Okinawa Attack: 14 October 1944
Northern Luzon and Formosa Attack: 13-14 October 1944
Luzon Attacks: 15 Oct 1944; 17-19, 5-16 November 1944; 19-25 November 1944
Visayas Attacks: 21 October 1944
Battle of Surigao Strait: 24-26 October 1944

1 Star/Okinawa Gunto Operation

World War II Victory Medal

Navy Occupation Service Medal
Asia: 14 September - 6 November 1945
Europe: 8 June - 31 July 1955;
1 August - 25 October 1955

China Service Medal (extended)
2 September - 9 October 1945

National Defense Service Medal (two awards)
(a) 9 February 1952 - 9 April 1952;
18 June 1954 - 27 July 1954
(b) 1 January 1961 -

Vietnam Service Medal
14 May - 15 June 1966;
8 July - 10 August 1966;
1-23 September 1966;
2-18 October 1966;
21 June - 13 July 1967;
29 July - 27 August 1967;
14 September - 12 October 1967;
31 October - 24 November 1967

Republic of Vietnam Campaign

Philippine Liberation Campaign

Republic of the Philippines Presidential Unit Citation

IV. Commanding Officers of the USS Intrepid 1943-1974

Captain T. L. Sprague	16 August 1943 - 28 March 1944
Commander Richard K. Gaines	28 March 1944 - 19 May 1944
Captain William D. Sample	19 May 1944 - 30 May 1944
Captain Joseph F. Bolger	30 May 1944 - 15 February 1945
Captain G. E. Short	15 February 1945 - 31 January 1946
Captain R. E. Blick	31 January 1946 - 11 April 1946
Captain H. G. Sanchez	11 April 1946 - 4 November 1946
Commander A. A. Giesser	4 November 1946 - 22 March 1947
Captain B. B. C. Lovett	9 February 1952 - 9 April 1952
Captain W. T. Easton	18 June 1954 - 26 May 1955
Captain G. L. Kohr	26 May 1955 - 15 August 1955
Captain P. P. Blackburn	15 August 1955 - 21 September 1956
Captain F. Massey	21 September 1956 - 4 October 1957
Captain J. H. Kuhl	4 October 1957 - 4 October 1958
Captain Paul Masterson	4 October 1958 - 30 August 1959
Captain Edward C. Outlaw	30 August 1959 - 8 September 1960
Captain C. S. Minter, Jr.	8 September 1960 - 24 May 1961
Captain J. L. Abbott, Jr.	24 May 1961 - 14 June 1962
Captain R. J. Morgan	14 June 1962 - 20 April 1963
Captain J. C. Lawrence	20 April 1963 - 31 April 1964
Captain J. G. Smith	31 April 1964 - 3 May 1965
Captain G. Macri	3 May 1965 - 15 July 1966
Captain J. W. Fair	15 July 1966 - 26 June 1967
Captain W. J. McVey	26 June 1967 - 22 Jun. 1968
Captain V.F. Kelley	22 Jun. 1968 - 1 Aug. 1969
Captain H.N. Moore, Jr.	1 Aug. 1969 - 1 Apr. 1970
Captain I.W. Linder	1 Apr. 1970 - 30 Apr. 1971
Captain C.S. Williams, Jr.	30 Apr. 1971 - 22 Dec. 1972
Captain R.H. Barker	22 Dec. 1972 - 10 Aug. 1973
Commander L.E. Levenson	10 Aug. 1973 - 22 Apr. 1974

V. Air Groups and Squadrons
U.S.S. *Intrepid*
1943-1972

September 1943 to January 1944: Air Group 8: VB-8; VF-8; VT-8, VF(N)-101.
January 1944 to February 1944: Air Group 6: VB-6; VF-6; VT-6; VF(N)-78.
June 1944 to August 1944: Air Group 19. (transport only).
August 1944 to November 1944: Air Group 18: VB-18; VF-18; VT-18; VF(N)-78.
Febuary 1945 to October 1945: Air Group 10: VB-10; VF-10; VT-10; VBF-10.
October 1945 to August 1946: Air Group 14.

1954: CVG-4: VF-22; VF-44; VF-173; VA-45; VC-12; VAH-7; HU-2; VC-4; VC-62; VC-33.
1955: CVG-4: VF-22; VF-44; VF-173; VA-45; VC-12; VAH-7; HU-2; VC-4; VC-62; VC-33.
1956: CVG-8: VF-82; VF-61; VA-83; VA-85; VFP-62; VC-12; VC-13;VAH-5; HU-2.
1957: N.A.T.O.: VF-71; VF-33; VA-66; VA-25; VAH-11; VFP-62.
1958: N.O.R. PAC: ATG-81; VA-42; VFP-62; VA-76; VF-41; VF-81; VAW-12; VA(AW)-33; HU-2.
1959: CVG-6: VA-66; VF-33; VA-65; VA-46; VF-74; VAW-12; VA(AW)-33; VFP-62; HU-2
1960: CVG-6: VA-66; VF-74; VA-65; VF-33; VA-76; VFP-62.
1961: CVG-6: VF-162; VF-33; VA-66; VA-65; VA-76; VAP-62; HU-2.
1962: VMA-225 (A-4 Squadron).
1963: CVSG-56: VS-24; VS-27; HS-3; VA(AW)-33.
1964: CVSG-56: VS-24; VS-27; HS-3; VA(AW)-33.
1966: CVW-10: VA-15; VA-95; VA-165; VA-176.
1967: CVW-10: VSF-3; VA-15; VA-34; VAP-63; VF-111; VA-145; VAW-121; VAW-33.
1968: CVW-10: VA-36; VA-106; VAW-121; VF-111; VFP-63; VA-66; VAQ-33.
1971: CVSG-56: HS-11; VA-45; AW-121; VS-24; VS-27; VS-31.
1972: CVSG-56: VS-27; VA-45; VS-24; VS-31; VAW-121; HS-11; VRC-40.

Intrepid's Average Time to Launch and Recover Planes*

Number of planes off per load	14
Time to launch one deck load	11minutes
Time to launch one from catapult	55 seconds
Time to launch one by fly away	32 seconds
Number to land	14
Time to land 14 planes	10 minutes
Time to land 1 plane	46.4 seconds

*Based on a 19 deck load launch for one day, with 14 planes per lauch for a total of 266 planes launched. Times of launch are based on non-battle action take offs. During emergencies these times are reduced. The first take-off is usually before down. The recovery is always completed by sunset. Launches after sunset are radar equipped night fighters. *Layout by A.F. Zollo*

VI. *Intrepid's* Aircraft from 1943

Make	Model	Year Introduced
Fighters:		
Grumman F6F-3	Hellcat	1943
Vought F4U-1	Corsair	1945
Grumman F9F-3	Panther	1948
Vought F7U-3	Cutlass	1951
Douglas F3D-2	Skynight	1951
McDonnell F3H-2	Demon	1955
McDonnell F2H-3	Banshee	1952
Grumman F9F-8	Cougar	1953
Douglas F4D-1	Skyray	1954
North American FJ-3	Fury	1955
Vought F8U-1	Crusader	1956
Grumman F11F-1	Tiger	1957
Attack Planes:		
Douglas SBD-5	Dauntless	1941
Grumman TBM-3	Avenger	1942
Curtiss SB2C-5	Helldiver	1944
Douglas AD-1	Skyraider	1946
North American AJ-2	Savage	1949
Douglas AD-4	Skyraider	1950
Douglas AD-5W (AE-1E)	Skyraider	1951
Douglas AD-6 (A-1H)	Skyraider	1952
Douglas A3D-2	Skywarrior	1954
Douglas A4D-2	Skyhawk	1956
Specialized Planes:		
Piasecki HUP-2		1950
Grumman TF-1	Trader	1955
Grumman S2F-2	Tracker	1956
Grumman WF-2	Tracer	1960
Sikorsky SH-3	Sea King	1961
Kaman UH-2C	Seasprite	1962

VII. Japanese Aircraft

Betty:	Twin engine medium/torpedo bomber (army & navy)
Emily	Four engine reconnaissance flying boat (navy)
Frances	Twin engine light/torpedo bomber (navy)
Frank	Single engine fighter (army)
George	Single engine carrier fighter (navy)
Hamp	Single engine carrier fighter (navy)
Jill	Single engine carrier torpedo bomber (navy)
Judy	Single engine carrier dive bomber (navy)
Kate	Single engine carrier torpedo bomber (navy)
Nell	Twin engine medium bomber (army and navy)
Pete	Single engine reconnaissance float plane (navy)
Tojo	Single engine fighter (army)
Tony	Single engine fighter (army & navy)
Val	Single engine carrier dive bomber (navy)
Zero (Zeke)	single engine carrier and land-based fighter (army & navy)

VIII. Nomenclature

Primary Combat Vessels

BB: Battleship
BM: Monitor
CV: Aircraft carrier
CVB: Aircraft carrier (large)
CVL: Aircraft carrier (small)
CVE: Aircraft carrier (escort)
CVS: Seaplane carrier
CC: Battle cruiser
CB: Large cruiser
CA: Heavy cruiser
CL: Light cruiser
DD: Destroyer
DL: Destroyer leader
DE: Destroyer escort
SS: Submarine
SM: Submarine minelayer

Mine Craft

CM: Minelalyer
CMc: Coastal minelayer
ACM: Auxiliary minelayer
DM: Light minelayer
DMS: High speed minesweeper
AM: Minesweeper
AMc: Coastal minesweeper
AMc (U): Coastal minesweeper (underwater locator)

Patrol Craft

PG: Gunboat
PR: River gunboat
PF: Frigate
PY: Yacht
PYc: Coastal yacht
PC: Submarine chaser (173 ft.)
PCC: Submarine chaser (control)
PCE: Escort
PCEC: Escort (control)
PCE(R): Escort (rescue)
PCS: Submarine chaser (136 ft.)
PCSC: Submarine chaser (control)
SC: Submarine chaser (110 ft.)
SCC: Submarine chaser (control)
PGM: Motor gunboat
PT: Motor torpedo boat
PTC: Motor boat, submarine chaser

Auxiliary Vessels

AB: Crane ship
AC: Collier
AD: Destroyer tender
AE: Ammunition ship
AF: Store ship
AG: Auxiliary (miscellaneous)
AGC: Amphibious force flagship
AGD: Seagoing dredger
AGL: Lighthouse tender.
AGP: Motor torpedo boat tender
AGS: Surveying ship
AGSc: Coastal surveying ship
AH: Hospital ship
AK: Cargo ship
AKA: Attack cargo ship
AKL: Light cargo ship
AKN: Net cargo ship
AKS: General stores-issue ship
AKV: Cargo ship and aircraft ferry
AL: Lightship
AN: Net laying ship
AO: Oiler
AOG: Gasoline tanker
AP: Transport
APA: Attack transport
APB: Barracks ship
APc: Coastal transport
APD: High speed transport
APH: Transport, evacuation of wounded
APL: Labor transport (non self-propelled)
APS: Transport submarine
APV: Transport and aircraft ferry
AR: Repair ship
ARB: Repair ship, battle damage
ARC: Cable ship
ARG: Repair ship, internal combustion engines
ARH: Heavy hull repair ship
ARL: Repair ship, landing craft
ARS: Salvage ship
ARS(D): Salvage lifting vessel
ARS(T): Salvage craft tender
ARV: Aircraft repair ship
AS: Submarine tender
ASR: Submarine rescue vessel
ATA: Ocean tug, auxiliary
ATF: Ocean tug, fleet
ATO: Ocean tug, old
ATR: Ocean tug, rescue
AV: Seaplane tender
AVC: Catapult lighter
AVD: Seaplane tender (destroyer)
AVP: Seaplane tender, small
AVR: Aircraft rescue vessel
AVS: Aviation supply ship
AW: Distilling ship
AWK: Water tanker
IX: Unclassified vessel

Landing Ships

LSD: Landing ship, dock
LSM: Landing ship, medium
LSM(R): Landing ship, medium (rocket)
LST: Landing ship, tank
LST(H): Landing ship, tank (evacuation)
LSV: Landing ship, vehicle
LCI(L): Landing craft, infantry (large)
LCI(G): Landing craft, infantry (gun)
LCI(M): Landing craft, infantry (mortar)
LCI(R): Landing craft, infantry (rocket)
LC(FF): Landing craft, infantry (flotilla flagship)
LCS(L): Landing craft, support (large)
LCT: Landing craft, tank
LCC: Landing craft, control
LCM: Landing craft, mechanized
LCP(L): Landing craft, personnel (large)
LCP(N): Landing craft, personnel (nested)
LCP(R): Landing craft, personnel (ramp)
LCR: Landing craft, rubber
LCV: Landing craft, vehicle
LCVP: Landing craft, vehicle-personnel
LVT: Landing vehicle, tracked

IX. U.S. Ship Losses, World War II

Battleships: (2 sunk)

Arizona (BB-39)	7 Dec. 1941, Pearl Harbor
Oklahoma (BB-37)	7 Dec. 1941, Pearl Harbor

Aircraft Carriers: (5 sunk)

Hornet (CV-8)	26 Oct. 1942, Santa Cruz Is
Lexington (CV-2)	8 May 1942, Coral Sea
Princeton (CVL-23)	4 Oct. 1944, Leyte Gulf
Wasp (CV-7)	15 Sep. 1942, Guadalcanal
Yorktown (CV-5)	4 Jun. 1942, Midway Is

Escort Carriers: (6 sunk)

Block Island (CVE-21)	29 May 1944, Canary Is.
Bismarck Sea (CVE-95)	21 Feb. 1945, Iwo Jima
Gambier Bay (CVE-73)	25 Oct. 1944, Samar Is.
Liscome Bay (CVE-56)	24 Nov. 1943, Gilbert Is.
Ommaney Bay (CVE-79)	4 Jan. 1945, Mindoro Is.
St. Lô (CVE-63)	25 Oct. 1944, Samar Is.

Cruisers: (10 sunk)

Astoria (CA-34)	9 Aug. 1942, Savo Is.
Atlanta (CL-51)	13 Nov. 1942, Guadalcanal
Chicago (CA-29)	30 Jan. 1943, Solomon Is.
Helena (CL-50)	6 Jul. 1943, Kula Gulf
Houston (CA-30)	1 Mar. 1942, Sunda Strait
Indianapolis (CA-35)	29 Jul. 1945, Philippine Sea
Juneau (CL-52)	13 Nov. 1942, Guadalcanal
Northampton (CA-26)	1 Dec. 1942, Tassafaronga
Quincy (CA-39)	9 Aug. 1942, Savo Is.
Vincennes (CA-44)	9 Aug. 1942, Savo Is.

Destroyers: (70 sunk)

Aaron Ward (DD-483)	7 Apr. 1943, Guadalcanal
Abner Reed (DD-526)	1 Nov. 1944, Leyte Is.
Barton (DD-599)	13 Nov. 1942, Guadalcanal
Beatty (DD-640)	6 Nov. 1943, Algeria
Benham (DD-387)	15 Nov. 1942, Guadalcanal
Blue (DD-387)	22 Aug. 1942, Guadalcanal
Borie (DD-215)	2 Nov. 1943, Azores
Bristol (DD-453)	12 Oct. 1943, Algiers
Brownson (DD-518)	26 Dec. 1943, New Britain Is.
Buck (DD-420)	9 Oct. 1943, Salerno
Bush (DD-529)	6 Apr. 1945, Okinawa
Callaghan (DD-792)	28 Jul. 1945, Okinawa
Chevalier (DD-451)	7 Oct. 1943, Solomon Is.
Calhoun (DD-801)	6 Apr. 1945, Okinawa
Cooper (DD-695)	3 Dec. 1944, Leyte Is.
Corry (DD-463)	6 Jun. 1944, Normandy, Fr.
Cushing (DD-376)	13 Nov. 1942, Guadalcanal
De Haven (DD-469)	1 Feb. 1943, Guadalcanal
Drexler (DD-741)	28 May 1945, Okinawa
Duncan (DD-485)	12 Oct. 1942, Guadalcanal
Edsall (DD-219)	1 Mar. 1942, Java
Glennon (DD-620)	10 Jun. 1944, Normandy, Fr.
Gwin (DD-433)	13 Jul. 1943, Kolombangara
Halligan (DD-584)	26 Mar. 1945, Okinawa
Hammann (DD-412)	6 Jun. 1942, Midway Is.
Henley (DD-391)	3 Oct. 1943, New Guinea
Hoel (DD-533)	25 Oct. 1944, Samar Is.
Hull (DD-350)	18 Dec. 1944, Luzon Is. (typhoon)
Ingraham (DD-444)	22 Aug. 1942, Nova Scotia
Jacob Jones (DD-130)	28 Feb. 1942, Cape May, N.J.
Jarvis (DD-393)	9 Aug. 1942, Guadalcanal
Johnston (DD-557)	25 Oct. 1944, Samar Is.
Laffey (DD-459)	13 Nov. 1942, Guadalcanal
Lansdale (DD-426)	20 Apr. 1944, Algeria
Leary (DD-158)	24 Dec. 1943, North Atlantic
Little (DD-803)	3 May 1945, Okinawa
Longshaw (DD-559)	18 May 1945, Okinawa
Luce (DD-522)	3 May 1945, Okinawa
Maddox (DD-622)	10 Jul. 1943, Sicily
Mahan (DD-364)	7 Dec. 1944, Leyte Is.
Mannert L. Abele (DD-733)	12 Apr. 1945, Okinawa
Meredith (DD-434)	15 Oct. 1942, Guadalcanal
Meredith (DD-726)	9 Jun. 1944, Normandy, Fr.
Monaghan (DD-354)	18 Dec. 1944, Luzon Is. (typhoon)
Monssen (DD-436)	13 Nov. 1942, Guadalcanal
Morrison (DD-560)	4 May 1945, Okinawa
O'Brien (DD-415)	19 Oct. 1942, Samoa
Parrott (DD-218)	2 May 1944, Norfolk, Va.
Peary (DD-226)	19 Feb. 1942, Port Darwin, Australia
Perkins (DD-377)	29 Nov. 1943, New Guinea
Pillsbury (DD-227)	1 Mar. 1942, Java
Pope (DD-225)	1 Mar. 1942, Java
Porter (DD-356)	26 Oct. 1943, Santa Cruz Is.
Preston (DD-379)	14 Nov. 1942, Guadalcanal
Pringle (DD-477)	16 Apr. 1945, Okinawa
Reid (DD-369)	11 Dec. 1944, Leyte Is.
Reuben James (DD-245)	31 Oct. 1941, Iceland;
Rowan (DD-405)	10 Sep. 1943, Salerno
Sims (DD-409)	7 May 1942, Coral Sea
Spence (DD-512)	18 Dec. 1944, Luzon Is.
Strong (DD-467)	5 Jul. 1943, Solomon Is.
Sturtevant (DD-240)	26 Apr. 1942, Florida
Truxtun (DD-229)	18 Feb. 1942, Newfoundland
Tucker (DD-374)	4 Aug. 1942, New Hebrides

Turner (DD-648)	3 Jan. 1944, New York
Twiggs (DD-591)	6 Jun. 1945, Okinawa
Walke (DD-416)	14 Nov. 1942, Guadalcanal
Warrington (DD-383)	13 Sep. 1944, Bahama Is.
William D. Porter (DD-579)	10 Jun. 1945, Okinawa
Worden (DD-352)	12 Jan. 1943, Aleutians

Note: Destroyers Cassin, Downes, and Shaw were sunk on 7 Dec. 1941 at Pearl Harbor but salvaged. Stewart was scuttled on 2 Mar. 1942 at Surabaya, salvaged by the Japanese and recovered after war's end. The following destroyers were all damaged beyond repair (kamikaze attacks) but not sunk: Evans, Haggard, Hugh W. Hadley, Hutchins, Leutze, Morris Newcombe, Shubrick, Thatcher. The Shaw was lost (again) by grounding at Leyte.

Destroyer Escorts: (11 sunk)

Eversole (DE-404)	28 Oct. 1944, Leyte Is.
Fechteler (DE-157)	5 may 1944, Algeria;
Fiske (DE-143)	2 Aug. 1944, Azores
Frederick C. Davis (DE-136)	24 Apr. 1945, N. Atlantic
Holder (DE-401)	11 Apr. 1944, Algiers
Leopold (DE-319)	10 Mar. 1944, Iceland
Oberrender (DE-344)	9 May 1945, Okinawa
Rich (DE-695)	8 Jun. 1944, Normandy, Fr.
Samuel B. Roberts (DE-413)	25 Oct. 1944, Samar Is.
Shelton (DE-407)	3 Oct. 1944, Morotai Is.
Underhill (DE-682)	24 Jul. 1945, Luzon Is.

Note: Destroyer Escort England was damaged (kamikaze) and not repaired.

Submarines: (52 sunk)

Albacore (SS-218)	7 Nov. 1944, Hokkaido
Amberjack (SS-219)	16 Feb. 1943, Rabaul
Argonaut (APS-1)	10 Jan. 1943, Rabaul
Barbel (SS-316)	4 Feb. 1945, Palawan Is.
Bonefish (SS-223)	18 Jun. 1945, Honshu
Bullhead (SS-332)	6 Aug. 1945, Bali
Capelin (SS-290)	Dec. 1943, Halmahera Is.
Cisco (SS-290)	28 Sep. 1943, Sulu Sea
Corvina (SS-226)	16 Nov. 1943, Truk
Darter (SS-227)	24 Oct. 1944, Palawan Is.
Dorado (SS-248)	12 Oct. 1943, Caribbean Sea
Escolar (SS-294)	Oct. 1944, Yellow Sea
Flier (SS-250)	13 Aug. 1944, Balabac Strait
Golet (SS-361)	14 Jun 1944, Honshu
Grampus (SS-207)	5 Mar. 1943, New Georgia
Grayback (SS-208)	26 Feb. 1944, East China Sea
Grayling (SS-209)	9 Sep. 1943, Luzon Is.
Grenadier (SS-210)	22 Apr. 1943, Penang
Growler (SS-215)	8 Nov. 1944, South China Sea
Grunion (SS-216)	30 Jul. 1942, Aleutians
Gudgeon (SS-211)	Apr. 1944, Mariana Is.
Harder (SS-257)	24 Aug. 1944, Caiman Point
Herring (SS-233)	1 Jun. 1944, Kuriles
Kete (SS-369)	Mar. 1945, Central Pacific
Lagarto (SS- 371)	3 May 1944, Gulf of Siam
Perch (SS-176)	3 Mar. 1942, Java
Pickerel (SS-177)	Apr. 1943, Honshu
Pompano (SS-181)	Aug. 1943, Honshu
Robaldo (SS-273)	26 Jul. 1944, Palawan
Runner (SS-275)	Jun. 1943, Kuriles
R-12 (SS-89)	12 Jun 1943, Florida
S-26 (SS-131)	24 Jan. 1942, Panama
S-27 (SS-132)	19 Jun. 1942, Aleutians
S-28 (SS-133)	4 Jul. 1944, Hawaii
S-36 (SS-141)	20 Jan. 1942, Makassar Strait
S-39 (SS-144)	14 Aug. 1942, Louisiade Archipelago
S-44 (SS-155)	7 Oct. 1943, Paramushiru
Scamp (SS-277)	6 Nov. 1944, Tokyo
Scorpion (SS-278)	Feb. 1944, Western Pacific
Sculpin (SS-191)	19 Nov. 1943, Truk
Sealion (SS-195)	10 Dec. 1941, Cavite
Seawolf (SS-197)	3 Oct 1944, Morotai Is.
Shark (SS-174)	11 Feb 1942, Celebes
Shark (SS-314)	24 Oct. 1944, Luzon Is.
Snook (SS-279)	Apr. 1945, Okinawa
Swordfish (SS-193)	Jan. 1945, Kyushu
Tang (SS-306)	24 Oct. 1944, Formosa
Trigger (SS- 237)	28 Mar1945, Nansei Shoto
Triton (SS-201)	15 Mar. 1943, Admiralty Is.
Trout (SS-202)	29 Feb. 1944, Okinawa
Tullibee (SS-284)	26 Mar. 1944, Palau
Wahoo (SS-238)	12 Oct. 1943, La Perouse Strait

Minelayers: (3 sunk)

Gamble (DM-15)	16 Jul. 1945, Saipan
Miantonomah (CM-10)	25 Sep. 1944, Le Harvre, Fr.
Montgomery (DM-17)	17 Oct. 1944, Palau

Note: Minelayers Aaron Ward and J. William Ditter were severely damaged and not repaired, April-June 1945.

Minesweepers: (15 sunk)

Bittern (AM-36)	10 Dec. 1941, Cavite
Finch (AM-9)	11 Apr. 1942, Corregidor
Minivet (AM-371)	29 Dec. 1945, Tsushima Strait
Osprey (AM 56)	5 Jun. 1944, Normandy, Fr.
Penguin (AM—33)	8 Dec. 1941, Guam

Portent (AM—106)	22 Jan. 1944, Anzio
Quail (AM—15)	6 May 1942, Corregidor
Salute (AM-294)	8 Jun. 1945, Borneo
Sentinel (AM-113)	12 Jul. 1943, Sicily
Skill (AM-115)	25 Sep. 1943, Isle of Capri
Skylark (AM-63)	28 Mar. 1945, Okinawa
Swallow (AML-65)	22 kApr. 1945, Okinawa
Swerve (AM-121)	9 Jul. 1944, Anzio
Tanager (AM-5)	4 May 1942, Corregidor
Tide (AM-125)	7 Jun. 1944, Normandy, Fr.

Highspeed Minesweepers: (6 sunk)

Emmons (DMS-22)	6 Apr. 1945, Okinawa
Hovey (DMS-11)	6 Jan. 1945, Lingayen Gulf
Long (DMS-12)	6 Jan. 1945, Lingayen Gulf
Palmer (DMS-5)	7 Jan. 1945, Lingayen Gulf
Perry (DMS-17)	13 Sep. 1944, Palau
Wasmuth (DMS-15)	29 Sep. 1942, Aleutians

Note: Minesweepers Butler, Forrest and Harding were severely damaged and not repaired, May—June 1945.

Patrol Vessels: (8 sunk)

Asheville (PG-21)	3 Mar. 1942, Java
Cythera (PY-26)	2 May 1942, Atlantic, NC
Erie (PG-50)	5 Dec. 1942, Willemstad
Luzon (PR-7)	6 May 1942, Corregidor
Mindanao (PR-8)	5 May 1942, Corregidor
Oahu (PR-6)	6 May 1942, Corregidor
Plymouth (PG-57)	5 Aug. 1943, Atlantic, NC
St. Augustine (PG-54)	6 Jan. 1944, Cape May, N.J.
Wake (PR-3)	7 Dec. 1941, captured at Shanghai

P.T. Boats: (69 sunk or lost)

69 Motor Torpedo Boats (P.T. boats) were lost in WWII due to a variety of actions (running aground, enemy submarine and surface action, mines, collision, weather, fire and explosion).

Submarine Tenders, Rescue Vessels: (3 sunk)

Canopus (AS-9)	10 Apr. 1942, Bataan
Macaw (ASR-11)	12 Feb. 1944, Midway Is.
Pigeon (ASR-6)	5 May 1942, Corregidor

Salvage Vessels: (3 sunk)

Extractor (ARS-15)	24 Jan. 1945, Philippine Sea
Redwing (ARS-4)	27 Jun. 1943, Bizerte
Rescuer (ARS-18)	1 Jan. 1943, Aleutians

Seaplane Tenders: (3 sunk)

Gannet (AVP-8)	7 Jun. 1942, Bermuda
Langley (AV-3)	27 Feb. 1942, Java
Thornton (AVD-11)	5 Apr. 1945, Okinawa

One of the aviation machinist mates at work in the shop 3 February, 1944. (National Archives Photo)

Oilers: (7 sunk)

Kanawha (AO-1)	7 Apr. 1943, Tulagi
Mississinewa (AO-59)	20 Nov. 1944, Ulithi
Neches (AO-5)	23 Jan. 1942, Hawaii
Neosho (AO-23)	11 Apr. 1942, Coral Sea
Pecos (AO-6)	1 Mar. 1942, Christmas Is.
Porcupine (IX-126)	30 Dec. 1944, Mindoro
Sheepscot (AOG-24)	5 Jun. 1945, Iwo Jima

Transports: (10 sunk)

Edward Rutledge (AP-52)	12 Nov. 1942, Morocco
George F. Elliot (AP-13)	8 Aug. 1942, Guadalcanal
Hugh L. Scott (AP-43)	12 Nov. 1942, Morocco
John Penn (APA-23)	13 Aug. 1943, Guadalcanal
Joseph Hewes (AP-50)	11 Nov. 1942, Morocco
Leedstown (AP-73)	9 Nov. 1942, Algiers
McCawley (APA-4)	30 Jun. 1943, Rendova
Susan B. Anthony (AP-72)	7 Jun. 1944, Normandy
Tasker H. Bliss (AP-42)	12 Nov. 1942, Morocco
Thomas Stone (AP-59)	25 Nov. 1942, Algiers

Highspeed Transports: (11 sunk);

Barry (APD-29)	21 Jun. 1945, Okinawa
Bates (APD-47)	25 May 1945, Okinawa
Belknap (APD-34)	11 Jan. 1945, Lingayan Gulf
Brooks (APD-10)	6 Jan. 1945, Lingayan Gulf
Colhoun (APD-2)	30 Aug. 1942, Guadalcanal
Dickerson (APD-21)	4 Apr. 1945, Okinawa
Gregory (APD-3)	5 Sep. 1942, Guadalcanal
Little (APD-4)	5 Sep. 1942, Guadalcanal
McKean (APD-5)	17 Nov. 1943, Bougainville
Noa (APD-24)	12 Sep. 1944, Palau
Ward (APD-16)	7 Dec. 1944, Leyte Is.

Cargo Vessels: (6 sunk)

Aludra (AK-72)	23 Jun. 1943, San Cristobal Is.
Deimos (AK-78)	23 Jun. 1943, San Cristobal Is.
Mount Hood (AE-11)	10 Nov. 1944, Admiralty Is.
Pollux (AKS-2)	18 Feb. 1942, Newfoundland
Pontiac (AF-20)	30 Jan. 1945, Halifax
Serpens (AK-97)	29 Jan. 1945, Guadalcanal

Tugs: (10 sunk)

ATR-15	19 Jun. 1944, Normandy Fr.
ATR-98	12 Apr. 1944, Azores
Genesee (AT-55)	5 May 1942, Corregidor
Grebe (AT-134)	5 Dec. 1942, Fiji
Napa (AT-32)	8 Apr. 1942, Bataan
Nauset (AT-89)	9 Sep. 1943, Salerno
Navajo (AT-64)	11 Sep. 1943, New Hebrides Is.
Partridge (ATO-138)	11 Jun. 1944, Normandy, Fr.
Seminole (AT-65)	25 Oct. 1942, Guadalcanal
Sonoma (ATO-12)	24 Oct. 1944, Leyte Is.

Intrepid takes injured man from destroyer. (National Archives Photo)

Auxiliaries: (7 sunk)

Ailanthus (AN-38)	26 feb. 1944, Aleutians
Atik	27 Mar. 1942, North Atlantic
Muskeget (AG-48)	9 Sep. 1942, North Atlantic
Niagara (AGP-1)	23 May 1943, San Cristobal Is.
Robert L. Barnes (AG-27)	8 Dec. 1941, Guam (captured)
Utah (AG-16)	7 Dec. 1941, Pearl Harbor
DCH-1 (IX 44)	28 Dec. 1941, Eastern Pacific

Coast Guard Vessels: (10 sunk)

Acacia (WAGL-200)	15 Mar. 1942, Haiti
Alex. Hamilton (WPG-34)	29 Jan. 1942, Iceland
Bedlow (WPC-128)	14 Sep. 1944, Cape Hatteras
Bodega (WYP-342)	20 Dec. 1943, Panama Canal
Dow (WYP-353)	14 Oct. 1943, Puerto Rico
Escanaba (WPG-77)	13 Jun. 1943, Greenland
Jackson (WPC-142)	14 Sep. 1944, Cape Hatteras
Magnolia (WAGL-231)	24 Aug. 1945, Mobile
Natsek (WYP-170)	17 Dec. 1942, Strait of Belle Isle;
Wilcox (WYP-333)	30 sep. 1943, Cape Hatteras

In addition to the above ships, the following 94 unnamed vessels were lost, none in 1942, a few in 1943, most in 1944 and a few in 1945.

LST (38); LSM (6); LSM(R) (3); APc (2); PC (5); SC (13); PGM (3); PE (1); YMS (20); CG (3).

X. Officers Aboard the U.S.S. Intrepid the Day of Her Commissioning

Captain
Sprague, Thomas Lamison

Commander
Gaines, Richard Kenna
Crissman, George Godfrey
Reynolds, Philip Shawcross
Lanman, Charles Burrows
Eyer, Alfred Warner
Karp, Henry Hugh

Lt. Commander
Fitzgerald, Charles Tuckermann
Wallace, Thomas Garrow
Fortune, James Hunter, Jr.
Garrison, Charles Ford
McLaughlin, Robert Bright
Thompson, Floyd Thomas
Lewis, William Edward
Marshall, Brewster Howison
McAlpin, John Volney
Matthiessen, Richard Peck
Arendt, Eric Harthan
Robinson, Ernest Kip
Kurtz, Gerald Independence

Lieutenant
Ramsey, John Breckinridge, Jr.
Hogan, Leo LaBrune
Adams, John Hughes
Heckman, Irvin James
Cranford, Wilson Hout
Kilpatrick, Macgregor
McElroy, Richard Stafford, Jr.
Fluckey, James Snowden
Hart, Thomas Patrick
Benchoff, Robert Johnston
Clark, James Thaddeus
Devens, Charles
Dunn, Arthur Autry, Jr.
Ziemen, James Saxton
Gribbel, John II
Allen, Erik Halway
Otteson, Kieth Vern
Combs, Paul Cortez
Fisher, Clarence Frazier, Jr.
Raymond, Frederick James
Morrill, James Francis
Duthie, John Milne
Hudson, William Jeffery
O'Connor, Thomas William
Davis, William Jesse, Jr.
White, John David
Caulfield, Edmund John
Wingate, Donald Farmer
Knight, Robert Clark
Cole, Howard
Moot, Richard Everett
Scheuble, Philip Arthur, Jr.
Short, Robert Earl
Miller, Charles Waldron
Cobb, William Jesse
Hennig, Geroge Carl
Herlihy, Timothy Francis
Schwabe, Leo Robert

Lieutenant (jg)
Guarnieri, Glen Joseph
Ruwwe, John William
De Clerque, Gerald
Lukens, William Parry
Gunther, Keith Edgar
Anderson, Robert Louis
Englebright, Harry Jackson
Clear, Thomas Leroy II
Ewing, John Sargeant
Whittaker, Joseph Augustin
Mountjoy, Leroy Irvine H.
Scisson, Sidney Eugene
Cox, Richard Wallace
Robuck, Charles Henry
Phillips, Stanley Edgar
Di Marzo, Donald Domenic
Feagin, Donald Ryan
Betten, Cornelius, Jr.
Shingleton, Gerald Coburn
Renfro, Joe Bert, Jr.
Witmer, Henry Herr
Wallace, John Clements
Machinsky, Joseph Edward
Davis, Thomas Crawley, Jr.
Watts, Albert Frazer
Wainright, Oscar Lyle, Jr.
Wagner, Roger Null
Peterson, Jinx
Reigher, Joseph Martin
Schreiber, Gerlad Harold
Hillas, Roland Alexander, Jr.
Wilson, Robert, Jr.
Ramsey, Paul James
Danner, Howard Ellis, Jr.
Yuschik, Theodore Fred
Lawrence, John Elmore

Ensign
Fletcher, Frederic Augustus
Martin, David Patrick
Lyness, Paul Iden
Halcrow, George Clark
Ballahtyne, John Niblack
Fortenberry, Earl Thomas
Shefferman, Shelton
Smith, Jay Howland
Fleetwood, William Everett
Segars, Henry Lee
Cohen, Emanuel Martin
Houston, Robert Charles
Kendall, George Pendleton
Carter, Robert Seriven
Taylor, Donald Francis
Frank, Howard Phillips
Barr, Lockwood Chappell
Rosset, Adolph
Small, Harold Charles
Morgan, Edmund Roy
Honour, Walter Whitaker
Seiler, Aubrey Richard
Shaw, Charley Binnet
McLeod, Charles Anderson
Krogstad, John Kristoffer
Murphy, Joseph Daniel
O' Leari, Valdo Peter
Bennett, William Myrt, Jr.
Clemmer, Leslie Curtis
Gill, Leon Burton, Jr.
Helgerson, Robert Aven
Simmons, Robert Louis
Johnson, Charles Leonard
Kline, Charles Tomlinson, Jr.
Peterson, Norman Charles
Duthie, James Ray, Jr.
Robbins, Arthur Wheelock
Boyle, Edward Hugh
Diaz, Anthony
Nuhn, John Alfred

Captain USMC
Hammond, Bruce Bennett

1st Lieutenant USMC
Keagy, Lowell Truscott

2nd Lieutenant USMC
Nickell, Louis "G"

Bo'sun
McDonald, Samuel Martin
White, Victor James
Johnson, Frank Elbridge

Gunner
Brown, Ralph James
Riter, John
Fitzgerald, James Bernard

Torpedo
Causey, Phines

Electrician
Marlowe, Delmas Edward
Montfort, Richard Lee
Norris, Charles Fletcher

C. Ra. Electrician
Harrer, Anthony John

C. Ra. Electrician
Russ, John Ernest, Jr.
Griebel, Raymond LeRoy

Machinist
Knoll, Jopseph Francis
Stepro, Alfred Woodrow
Schlemmer, George Steven
Herman, Otto
Brewer, Richard Lester
Gay, Homer Alison
Bollheimer, Joseph Carl
Blaylock, John Roscoe
Morelock, Ober Berl

Carpenter
Diggs, Walter
Cox, Homer Lawrence
Hitt, Samuel Luther
Priest, Dean William

Clerk
Neally, Rochester Hampton
Garner, Syvester Lee, Jr.

Ch. P. Clerk
Gosting, George Guerin
Carver, Oro Addison

Act. P. Clerk
Phipps, Albert Clayton
Lunde, Norman Clifford

Pharmacist
Maddox, Theo Posey, Jr.

U.S. ARMY PHILIPPINE ISLANDS CTF 77
General of the Army — Douglas MacArthur
Admiral — Thomas C. Kincaid